T5-ARJ-775

Lecture Notes in Biomathematics

Managing Editor: S. Levin

80

Donald House

Depth Perception in Frogs and Toads

A Study in Neural Computing

Springer-Verlag

New York Berlin Heidelberg London Paris Tokyo Hong Kong

Author

Donald House
Department of Computer Science
Williams College
Williamstown, MA 01267, USA

Mathematics Subject Classification (1980): 68 T XX, 92 A 18

CR Subject Classifications (1982): J.3; I.2.m

ISBN 0-387-97157-2 Springer-Verlag New York Berlin Heidelberg Tokyo
ISBN 3-540-97157-2 Springer-Verlag Berlin Heidelberg New York Tokyo

Printed in Germany

Printing and binding: Beltz Offsetdruck, Hemsbach/Bergstr.
2849/3140-543210 – Printed on acid-free paper

Depth Perception in Frogs and Toads -- A Study in Neural Computing provides a
mprehensive exploration of the phenomenon of depth perception in frogs and toads, as
en from a neuro-computational point of view. Perhaps the most important feature of the
ok is the development and presentation of two neurally realizable depth perception
orithms that utilize both monocular and binocular depth cues in a cooperative fashion.
e of these algorithms is specialized for computation of depth maps for navigation, and
other for the selection and localization of a single prey for prey catching. The book is
o unique in that it thoroughly reviews the known neuroanatomical, neurophysiological
d behavioral data, and then synthesizes, organizes and interprets that information to
plain a complex sensory-motor task.

The book should be of special interest to that segment of the neural computing
mmunity interested in understanding natural neurocomputational structures, and will be
particular interest to those working in perception and sensory-motor coordination. It
ll also be of interest to neuroscientists interested in exploring the complex interactions
tween the neural substrates that underly perception and behavior.

nald H. House
illiamstown
ly 14, 1989

Contents

1

Introduction

In this study, we present a sequence of models exploring neural mechanisms by which frogs and toads may extract depth information from visual data. Although the models are constructed to be true, as far as possible, to what is known regarding depth perception in frogs and toads, they are also of more general interest. In particular, when they were developed they were the first depth models to exploit both binocular and monocular depth cues in a cooperative fashion.

Our work differs from most current work in depth perception in several ways. Of course, the first of these is that we use frogs and toads, rather than mammals, as our study system. Second, we are particularly concerned with the ways in which depth information from multiple cue sources can be integrated. Finally, our point of view is that perception is action-oriented. This has required us to pay careful attention to behavioral as well as anatomical and physiological data in our search for plausible neural algorithms.

Our analysis suggests that frogs and toads do not use a single means for determining depth from visual input but, instead, use at least two different processes, one specialized for navigation and the other for prey catching. We present models of these two processes and outline features of an accommodation control system that will support them both. Our model supporting navigation consists of two cooperative depth mapping processes, with one process receiving accommodation cues and the other receiving binocular disparity cues. By sharing information, the two processes are able to build a single depth map, whose accuracy is governed by binocularity. Our second model is of a process best suited for prey catching. Instead of building a depth map, it determines the spatial location of a single object by coupling prey-selection and lens-accommodation processes within a feedback loop. Information derived from prey selection supplies a setpoint for accommodation. In turn, lens accommodation modifies the visual input and affects the prey selection process. This scheme automatically corrects spurious binocular matches.

Our purpose in creating these models is twofold and involves two disparate disciplines. First, the models are constrained to conform to existing knowledge of the visuomotor system of frogs and toads and are aimed at improving our understanding of visuomotor coordination in these animals. Second, by constructing and experimenting with computer simulations of the models, we have taken a first step in the development and testing of sensory-motor control algorithms for artificial systems.

Experiments done with the computer simulations allowed us to go beyond mere plausibility arguments by providing concrete demonstrations of the efficacy of the models. They allowed us to analyze the sensitivity of the models to variations in parameter settings and in environmental configurations. This analysis, in turn, led to many concrete predictions about animal behavior that are now open for verification by experimentalists. Further, the computer simulations represent operating algorithms that can be implemented in robotic systems. Experiments with these simulations provide important data concerning the tuning and expected performance characteristics of these algorithms.

We approached this study of depth perception with the view that, when addressing the problems of sensory-motor coordination, there are no boundaries between artificial and natural sensory-motor systems. The study of natural systems, which are so successful at performing sensory-motor tasks that are difficult to implement in machines, holds promise of providing elegant solutions to practical problems in robotics. Conversely, when a solution to a sensory-motor problem has proven successful in an artificial system, it becomes a rich source of analogy in the study of natural systems.

Sensory processes and motor processes cannot and should not be put into separate categories. The purpose of sensation is to adjust action. Action in turn affects perception. Action and perception are two inseparable parts of a closed cycle [Neisser, 1976; Arbib, 1981]. If depth discrimination is embedded within a cycle of action and perception, then it cannot be viewed simply as an information-organizing process separate from action. If, for instance, a toad exhibits size-constancy in its preference for prey (i.e. it appears to make judgements about visually presented objects that reflect the object's real size rather than subsumed visual angle) then do we assume that the toad estimates the distance of the prey and then determines its size, or do we assume that a side effect of the toad's prey acquistion strategy is that prey of a certain size are preferred?

Our point of view does not allow us to treat depth perception as a passive process, concerned only with assigning a depth value to each point in visual space. Instead, it must be viewed as part of a dynamic process by which an animal (or machine or human) achieves a particular goal. If the goal is navigation through the environment, then what is required is quite different from what is required if the goal is to catch a passing fly. Depth perception, to the extent that it is an identifiable function of the visual system, may take as many different forms as there are distinctly different functions that require its use.

Frogs and toads were chosen for this study for several reasons. First, they exhibit ballistic behavior. Their prey-catching sequence consists of periods of apparent inactivity followed by a series of movements that seem to be preplanned. Changes in the environment after initiation of overt activity do not seem to affect the way in which that sequence is carried out. Once a toad begins to stalk and strike at a prey object, it will not stop even if the prey target is artifically removed; the sequence, including swallowing and wiping the mouth after the strike, is uninterrupted [Ewert, 1980]. Ballistic behavior is remarkably akin to the actions of preprogrammed robots. A second reason for choosing frogs and toads is that they are vertebrates, but have relatively simple brains. Therefore, it may be possible to draw parallels to visual functioning in higher vertebrates without having to confront their greater complexity [Scalia and Fite, 1974]. Finally, frogs and toads exhibit highly developed depth vision but also appear to utilize fairly simple rules when using depth information to coordinate their behavior [Collett, 1982]. These rules employ only readily available visual cues, and thus they are of the sort which could conceivably be implemented in a robotic system.

We have carefully examined data from experimentalists and hope that, in turn, our models will be the subject of experimental testing. We also hope that our models will prove useful in the design of control systems for robots, even if results of experients with frogs and toads prove to refute some of our speculations.

Acknowledgements: The research reported in this book was initially supported by NIH grant NS14971 03–05, M. A. Arbib principle investigator. Later work and preparation of the manuscript were partially supported by Williams College. The author would like to extend special thanks to Michael Arbib for his guidance throughout this project, to Andrew Barto and Thomas Collett for their scientific guidance, and to Ann Ferguson Nauman for her expert and insightful editing.

2

Modeling Frog and Toad Depth Perception

ABSTRACT In this chapter we review the reasons why previous theoretical models do not provide an adequate explanation for the depth perception process in frogs and toads. We begin with a survey of earlier depth models and a review of experimental evidence from behavioral, anatomical, and physiological experiments relating to depth perception in frogs and toads. We conclude by contrasting the assumptions made in the models with observations about the animals. We note that all of the previous depth models based on stereopsis are restricted to the consideration of binocular matching on a pair of static images, and that their purpose is to produce a depth-mapping from the image pair. These models also depend upon assumptions concerning vergence and image granularity. We show that these restrictions and assumptions are not applicable to the depth resolution problem in frogs and toads.

2.1 Previous Models of Depth Perception

Strategies for obtaining depth information from optical data fall into three main categories: 1) optic flow algorithms, 2) autofocus algorithms, and 3) stereoscopic algorithms. Optic flow algorithms utilize a time-sequence of images from a single sensor (camera or eye) together with sensor movement data to reconstruct depth from parallax between consecutively scanned images. Some of the more recent work on optic flow algorithms is described in Prager and Arbib [1983], and Lawton [1984]. Autofocus algorithms attempt to adjust the lens of the sensor to maximize some measure of image focus. In this sense they are not depth algorithms at all, but once the lens is adjusted depth can be inferred from its setting. A short review of the various autofocus algorithms is contained in Ligthart and Groen [1982]. Selker [1982] describes a hardware implementation of such an autofocusing system. Stereoscopic algorithms address the problem of obtaining depth from binocular images by computing disparities between matching regions on the two images.

Neither the optic flow nor the autofocus algorithms were deemed appropriate to the study of depth perception in frogs and toads. The optic flow algorithms all entail the use of a moving sensor, but frogs and toads are able to determine depth while remaining stationary and exhibit no tracking eye movements [Autrum, 1959; Ewert, 1980]. Existing autofocus algorithms all use a global measure of image focus to adjust the lens. Therefore, they are most useful with sensors having a narrow field of view. However, the monocular visual field of frogs and toads exceeds 180° [Fite and Scalia, 1976].

The stereoscopic algorithms are also not directly applicable to the depth perception problem in frogs and toads, but the reasons for their unsuitability are more subtle. The remainder of this chapter provides an analysis of these subtleties.

Studies of human depth discrimination in random-dot stereograms have provided most of the incentive for research in stereoscopic depth vision. Julesz [1971] showed that humans could successfully fuse a stereo pair of random-dot patterns so that if a central rectangle in one image (selected arbitrarily, and not defined by edges in the image) were shifted with respect to that in the other image the observer would see this central pattern as either hovering above or sunken into the surrounding pattern. This ability to binocularly match regions of a pair of images without the use of high-level monocular features shows that pattern recognition is not necessary for binocular depth resolution.

Nelson [1975] developed a neurophysiologically grounded computational model that explained Julesz' results. Dev [1975] and later Marr and T. Poggio [1976] developed alternative models following the principles outlined by Nelson. Their models are based on cooperation and competition in spatially distributed arrays of neuron-like elements. In these models a single disparity (or depth) is assigned to each retinal position by choosing the most likely estimate at that position. In both of these models similar assumptions are made about the input image pair. The most important assumption is that of continuity, the observation that objects in the world tend to have continuous surfaces which vary only gradually in depth except at object boundaries. Because of continuity, most neighboring image points share similar disparities. Thus, a reasonable computational scheme is to weigh a depth estimate more heavily when it is similar to estimates at neighboring image sites. The algorithms are made more efficient if it is assumed that the eyes are verged upon a spatial point near the image region that is being analyzed. This assumption assures that the range of disparities is small and that, in the absence of other cues, a smaller disparity is more likely than a larger one.

Hirai and Fukushima [1978] extended these earlier cooperative/competitive schemes by introducing explicit cues from vergence to assist in disambiguation of binocular matching. In addition to using neighboring estimates in their weighting scheme, they also use a global matching process which helps to assure that 1) a point in one image is matched by only one point in the other image, and 2) that the ordering of the matches is consistent.

Trehub [1978] also extended the earlier cooperative/competitive algorithms. He proposed that performance could be improved by using results of a local pattern coorelation, in addition to biasing from neighboring estimates, to assure proper binocular matching. His scheme requires the additional assumption that a fine-grained pattern of stimulation be available at each local matçhing site.

Amari and Arbib [1977] present a theoretical study of the general class of cooperative/competitive "selection" algorithms. Their work includes several theorems which relate parameter settings to model equilibria.

More recently, modelers have exploited spatial-frequency tuned channels to guide the binocular matching process in a hierarchical way. Models by both Marr and T. Poggio [1979], and Frisby and Mayhew [1980] match features of spatial-frequency band-passed versions of the images, so that the matches obtained from the low-frequency versions of the image can be used to bias the matching process in the higher-frequency components. The models differ in that Frisby and Mayhew follow previous models in using a cooperative scheme whereas Marr and T. Poggio do not.

The common purpose of all of these depth models is to produce a depth mapping of the image under consideration. To do this, the models employ only binocular depth cues and resort only to the information contained in one static stereo-pair of images. An assumption (stated or otherwise) that is made in all of these models is that binocular convergence has succeeded in producing images in which the range of binocular disparities is both small and centered about zero. It is also implicitly assumed that the images are clearly focused. Finally, all of these models depend upon a fine-grained pattern of visual stimulation being available to the matching algorithm. In order to evaluate the applicability of these assumptions to the depth perception system of frogs and toads we will first examine selected background material on the visual system of these animals.

2.2 Depth Perception in Frogs and Toads

In this section we present data from behavioral, anatomical, and physiological studies that relate to the perception of depth by frogs and toads. Behavioral results demonstrate how depth information is utilized, and indicate that depth information comes from both binocular and monocular cue sources. Anatomical and physiological findings indicate those brain structures that seem most likely to be involved in the perception of depth. Our review traces the flow of visual information from the retinas to the brain-stem motor centers. We also present the results of several studies that correlate the anatomical and physiological units with behavior.

2.2.1 THE ROLE OF DEPTH PERCEPTION IN DETOUR BEHAVIOR

When they are involved in catching prey or avoiding predators, frogs and toads use depth information in surprisingly sophisticated ways. For example, when approaching a prey located behind a barrier, frogs and toads exhibit size constancy for both prey and barrier [Lock and Collett, 1979; Ingle, 1976]. They also demonstrate the ability to measure the distance between barrier and prey [Lock and Collett, 1980], and they can remember the position of a prey and orient towards that position even after the prey is no longer visible [Collett, 1982].

The choices that toads make in detouring around a barrier to reach prey give clues as to how they utilize depth information. Lock and Collett [1979, 1980] demonstrated that, when confronted with a paling-fence barrier between it and its prey, a toad chooses either to detour around one of the fence ends, to push through the fence in an attempt to reach the prey directly, or to ignore the prey. Later, Collett [1982] showed that, in choosing among these three options, toads make decisions that depend upon barrier position and length, distance and position of prey relative to the barrier, and the placement and width of gaps in the barrier. In all cases, these judgements of length, width, and distance appear to be based on estimates of the actual dimension and not on a simple measurement of subsumed visual angle.

Toads are able to locate simultaneously two barrier surfaces in depth. Fig. 2.1 shows the results of several of Collett's experiments where he presented toads with various barrier/prey configurations. Fig. 2.1a shows a toad faced with a distant unbroken fence between it and its prey. In this case, 75% of the approaches were aimed at one of the fence ends, with only 25% aimed directly at the prey. Fig. 2.1b shows a nearer fence with a central gap. Here, approaches were aimed almost entirely at the gap. However, when two fences, similar to those above, were presented simultaneously, as in Fig. 2.1c, the toads showed a response that was unlike either response to the individual fences. Instead, it was more nearly an average of the two responses, with about 50% of the approaches being directed toward a fence end and the other 50% directed at the gap in the front fence. One explanation for these results is that the toad actually sees the two fences in separate depth planes. Collett showed that toads can measure the distance between two fences by running the same experiment but this time varying the distance between the two fences. He found that the percentage of approaches aimed at the frontal gap increased as the distance between the fences increased.

The ability to discriminate the distance between fences is not based upon measurement of the difference in angular distance between the fenceposts. To prove this, Collett tried several configurations involving a single fence with a central segment whose palings were more closely placed than those

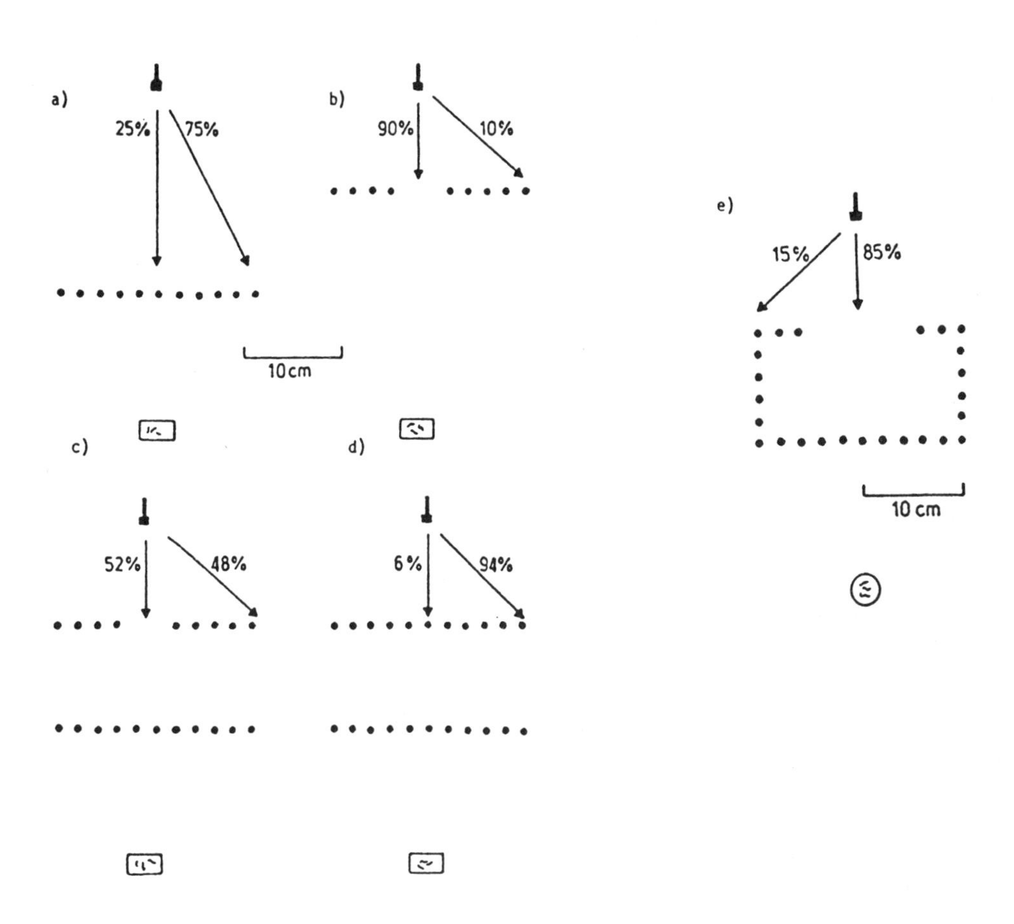

FIGURE 2.1. Prey/Barrier Experiments
Rows of dots represent paling fence barriers, large rectangles and circle indicate prey position, and inverted T's represent the position and orientation of a toad. Percentages indicate the number of trials on which the animals elected to head directly toward the prey versus the number on which they elected to detour around either of the fence ends. Reprinted by permission from Collett [1982].

in the rest of the fence. Results were similar to those obtained when all fenceposts were evenly spaced. Approaches still either detoured around the fence or were directed toward the prey. The toads never oriented toward the position where the spacing between the fenceposts changed.

Toads are able to measure the distance between a barrier and a prey object. Lock and Collett [1980] showed that when a toad is confronted with a prey object behind a fence-like barrier, the distance between the toad and the fence does not affect the percentage of approaches that detour around the fence. As long as the distance between the prey and the fence is held constant, it does not matter how far the configuration is from the toad. However, the distance between the prey and the fence has a major effect upon approach direction preference. Moving the prey closer to the fence increases the number of direct approaches, whereas moving the prey farther from the fence increases the number of detours. Collett [1982] showed that this effect is maintained even when there are two fences and the prey is behind the rear fence.

During prey-catching, toads plan a route to their prey which they can follow whether or not the prey stays in view during the approach. Fig. 2.2 depicts another experiment by Collett [1982] that demonstrates this fact. In this experiment, a paling fence was interposed between the toad and its prey. Behind the fence was a T-shaped opaque barrier that allowed the toad to view the prey only when the toad's long axis was aligned with the channel forming the leg of the T. Once the toad began its detour approach, the prey was hidden by the barrier. Nevertheless, turns around the barrier ends were directed toward the vicinity of the prey. If the toad elected to continue its approach after rounding the barrier, it tended to head toward the position of the prey. These data make clear the toad's ability to extract depth information from its visual world, to maintain a short-term memory of this depth information, and to integrate this memory with some notion of its own body movement.

2.2.2 Monocular and binocular depth cues

Frogs and toads use both monocular and binocular cues to determine depth. When both types of cue are available, the binocular cues are preferred. Ingle [1976] demonstrated that monocular frogs are able to snap accurately at prey located throughout the visual field of their intact eye, although there is some tendency to undershoot targets placed contralateral to the midline. Fig. 2.3 using measurements from cine recordings, illustrates this result. Line B (binocular frogs) slightly overlaps the stimulus across the entire rostral binocular field, while line M (monocular frogs) falls short of the target for locations 15°–35° from the midline in the contralateral visual field. Errors within the ipsilateral visual field were not significantly greater for monocular animals than for binocular animals. Thus it may be concluded that monocular information is sufficient for depth perception in

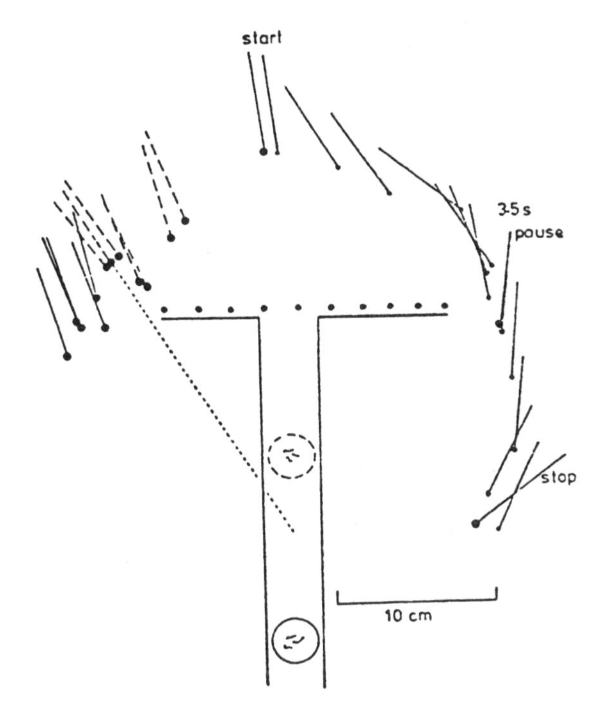

FIGURE 2.2. Toad Trajectories
Solid lines to the right indicate the orientation of the body axis of a toad and its snout position (dots) at intervals along its path toward the prey, enclosed within the solid circle. Solid lines to the left show the orientation of the toad's body axis, for several trials, during its pause at the fence end. Dashed lines are similar but for prey positioned within the dashed circle. Reprinted by permission from Collett [1982].

frogs. The presence of significant error only in the periphery, where lens resolving power is least, led Ingle to hypothesize that lens accommodation provides the monocular depth cues.

Collett [1977] found that toads (*Bufo marinus*) are also able to estimate depth monocularly. Fig. 2.4b shows that both binocular and monocular toads are able to judge prey distance accurately. Snapping distance for nearly all trials showed the overshoot illustrated by Ingle in Fig. 2.3. Collett confirmed Ingle's hypothesis on accommodation by showing that a concave lens placed in front of the eye of a monocular toad will cause it to undershoot prey (Fig. 2.4c). However, Collett also showed that binocular toads utilize binocular depth cues almost exclusively for estimating the depth of targets in the binocular field. He found that lenses have only a weak effect on depth estimation in binocular toads (Fig. 2.4c) but that base-out prisms placed in front of their eyes produce significant undershooting. Fig. 2.4a shows that the prism-induced angular shifts in retinal position of the physical object result in its being perceived as nearer than it actually

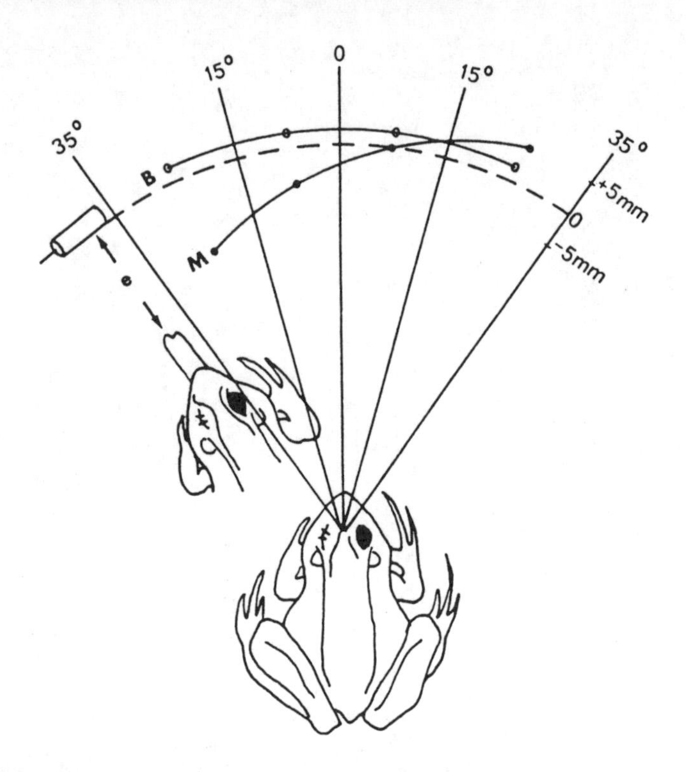

FIGURE 2.3. Snapping Errors After Monocular Blinding
Solid lines mark the deviation of the tip of the wet tongue mark from the near edge of the cylindrical target (dashed line). Line B shows the mean tongue extension for binocular frogs and line M for monocular frogs. Reprinted by permission from Ingle [1976].

is (virtual object). In Fig. 2.4d, lines show the snapping distance predicted as a result of the effect of three prismatic strengths on visual parallax. Distortion of depth estimation was progressively greater with increasing prismatic strength, confirming the dominance of binocular depth cues over monocular cues. Collett was able to closely match his experimental data with a formula that gives a 94% weighting to binocular cues and only 6% to monocular cues.

A more recent series of experiments, carried out in Roth's laboratory, offers strong confirmation of the results of Collett and Ingle. Jordan et al. [1980] tested the depth resolving ability of toads (*Bufo bufo*) after topical application of drugs to the eye to alter tension in the lens accommodation muscles. They treated binocular toads with atropine (a mydriatic, or pupil dilator, and thus a relaxant of lens accommodation muscles) and with physostigmine (a miotic, or pupil contractor, and thus a contractor of lens accommodation muscles). They found that treatment of normal binocular toads with these two drugs had only a small effect on their perception of

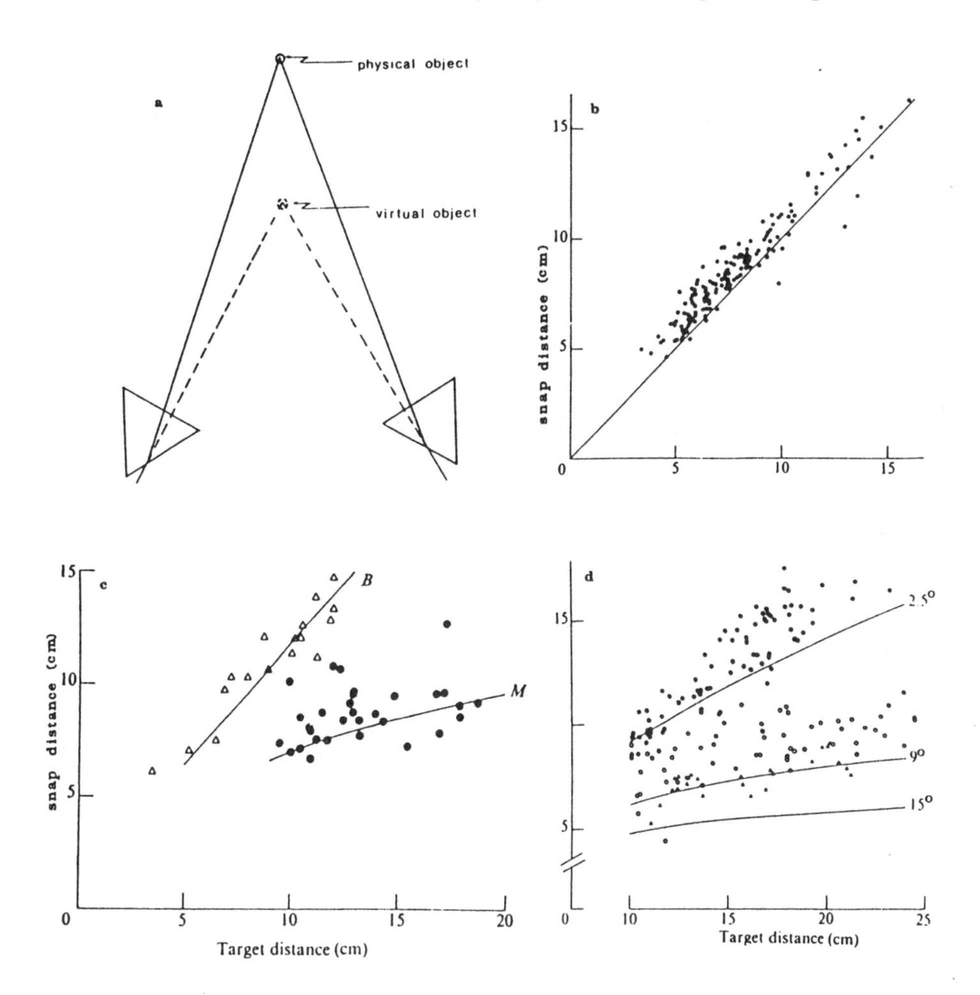

FIGURE 2.4. Effect of Prisms and Lenses on Depth Perception
(a) Error in distance estimate caused by base-out prisms. (b) The oblique line represents the exact distance of the target from the animal. Filled circles show the results of trials with binocular toads and the open circles with monocular toads. (c) The effect of concave lenses on snapping distance. Open triangles represent trials with binocular animals, filled circles trials with monocular animals. Line B shows the snapping distance expected (with a small correction for the prismatic effect of the lenses) if the change in focal length due to the lenses has no effect upon depth estimation. Line M shows the snapping distance predicted by the change in focal length. (d) The effect of prisms of three different strengths on snapping distance. Filled circles represent trials with 2.5% prisms, open circles with 9% prisms, and triangles with 15% prisms. Reprinted by permission from Collett [1977].

the depth of prey. However, monocular toads tended to severely undershoot prey after treatment with the muscle relaxant and to overshoot after treatment with the muscle contractor.

2.3 Anatomy and Physiology

2.3.1 The eyes

In both frogs and toads, the eyes are outward-facing but are structured to provide vision over nearly the entire superior hemisphere. The inferior hemisphere is also well covered, particularly in the frontal area. Fig. 2.5 (from Fite and Scalia [1976]) shows this field for both *Rana pipiens* and *Rana catesbeiana*. A significant area of binocular overlap is apparent. For instance, the frontal binocular field covers more than 60° in *Rana pipiens*. Grobstein et al. [1980] report similar findings.

Frogs and toads do not make vergent eye movements, and eye movements are not made to fixate or track objects in the visual field [Autrum, 1959; Ewert, 1980]. The small eye movements that do occur, such as those made during opto-kinetic nystagmus [Montgomery et al., 1982], are probably related to image stabilization. Without the ability to binocularly converge the eyes, there is no means by which a selected portion of the image can be fixated at a known point on the two retinas. Consistent with their lack of vergence, the eyes of frogs and toads have no distinct foveal area. Instead, photoreceptors and the ganglion cells contributing to the optic nerve are distributed relatively evenly (compared with higher vertebrates) over the retinal surface, as shown in Fig. 2.6 (from Fite and Scalia [1976]).

2.3.2 Optic nerve projection sites

The optic nerves cross at the chiasm, sending most fibers to contralateral diencephalic and mesencephalic brain regions [Scalia and Fite, 1974]. Dense retinotopically organized projections extend to the optic tectum with smaller projections to several thalamic nuclei and to the basal optic nucleus. These projections are shown in Fig. 2.7. The tectum is the major retinal projection site. Projections to tectum are monocular, coming almost entirely from the contralateral retina. However, within the thalamus, the corpus geniculatum, nucleus of Bellonci, posterior thalamic nucleus, and uncinate pretectal nucleus all receive binocular input. Ipsilateral projections from the binocular visual field to these thalamic nuclei are as dense as the contralateral projections [Fite and Scalia, 1976]. Projections to sites in the dorsal anterior thalamus and the basal optic nucleus are monocular. Beneath the most anterior portion of the tectum, and thus not shown in Fig. 2.7, is the large-celled pretectal nucleus which receives a monocular projection.

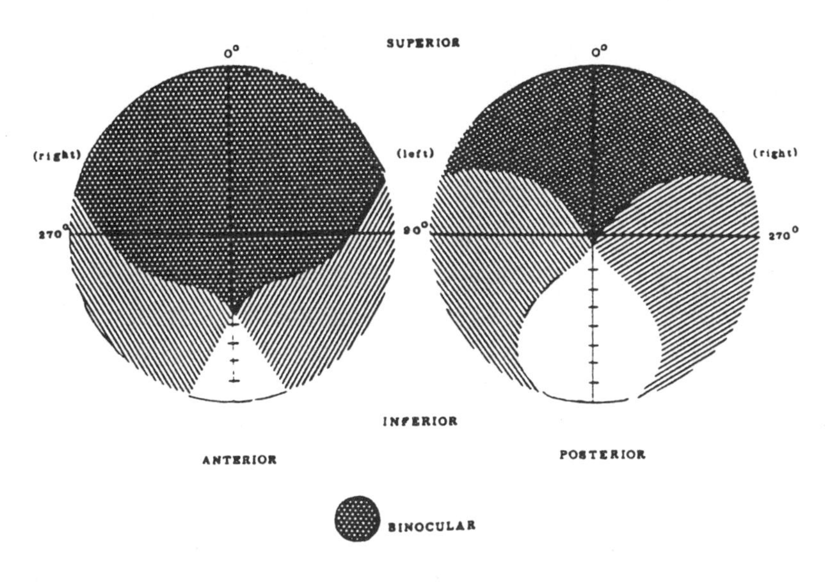

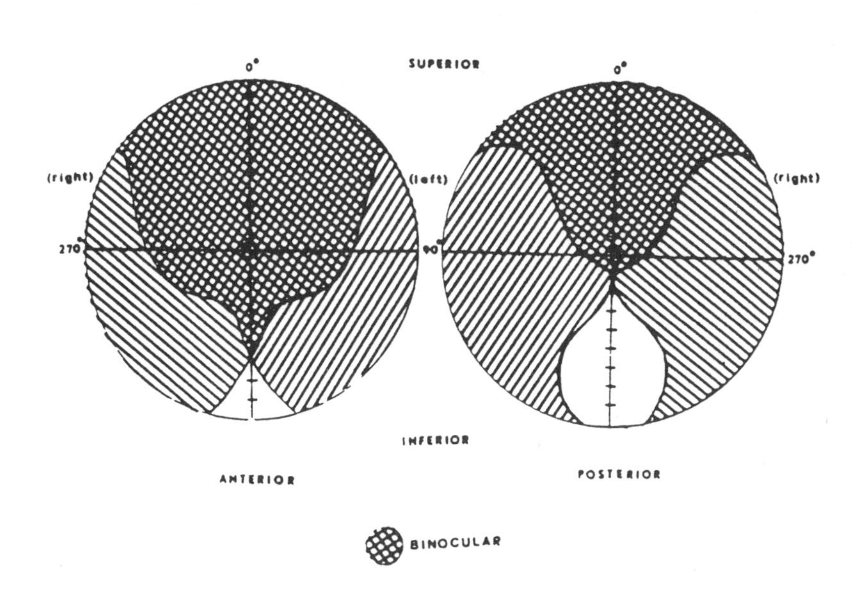

FIGURE 2.5. Visual Field Maps in Frog
Perimetric maps of anterior and posterior fields of view as measured for *Rana pipiens* and *Rana catesbeiana* showing the visual field of the right and left eye and the region of binocular overlap. Reprinted by permission from Fite and Scalia [1976].

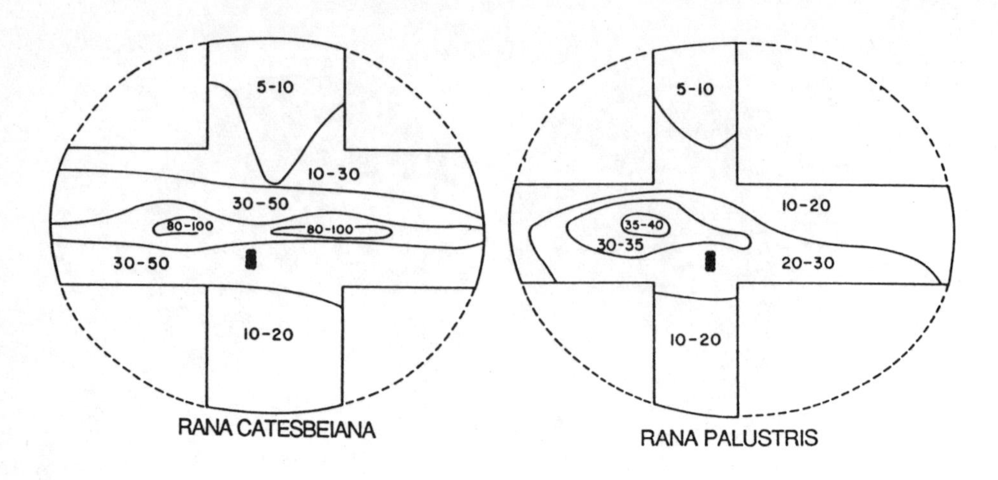

FIGURE 2.6. Retinal Receptor Density Maps
Reprinted by permission from Fite and Scalia [1976].

Although the optic nerve projection sites in thalamus are called "nuclei", they would more accurately be labeled "retinal recipient neuropil." They are actually regions of dense synaptic contact with the optic nerve. Thalamic cell bodies send dendrites into the nuclei to contact retinal ganglion cell axons. Few axons of thalamic neurons extend into this region [Scalia and Fite, 1974].

Optic nerve terminations in tectum are also far from the cell bodies of neurons with which they form synapses [Grüsser and Grüsser-Cornehls, 1976]. These terminals all lie in the relatively cell-free superficial layers of the tectum. Apical dendrites from cells in deeper tectal layers project into the superficial layers. Here they intertwine with afferent fibers to form a glomerular structure.

2.3.3 Nucleus isthmi — a source of binocular input to tectum

In addition to receiving direct monocular projections from the contralateral eye, each hemisphere of the optic tectum receives an indirect projection from the ipsilateral eye. This ipsilateral projection is visuotopically organized and in register with the contralateral projection. Thus, that portion of the tectum devoted to the binocular visual field receives a complete binocular projection. The notion of two retinotopic projections being in register implies the existence of a surface in space, any point on which maps through both eyes to the same point on the neural surface. In animals exhibiting vergence, this surface is called the horopter and, being the surface of zero disparity, always contains the fixation point. Frogs and toads do not exhibit vergence but nevertheless there is such a surface. In

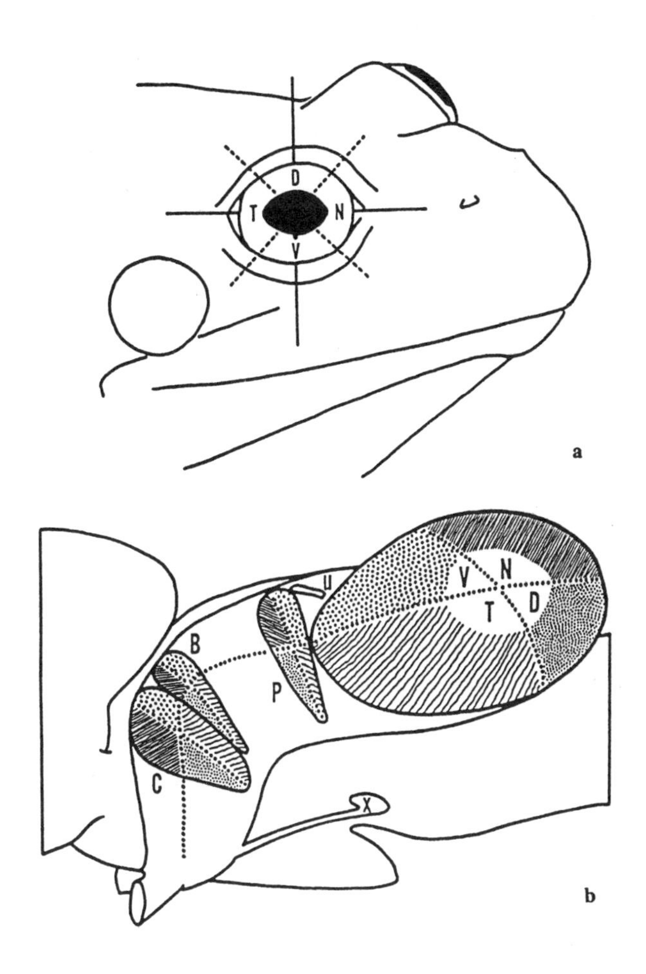

FIGURE 2.7. Optic Nerve Projection Sites
(a) Labeling convention for the four quadrants of the eye: temporal T (anterior visual field), dorsal D (inferior visual field), nasal N (posterior visual field), and ventral V (superior visual field). (b) Flattened sketch showing the left side of the brain and the major optic nerve projection sites. C is the corpus geniculatum, B the nucleus of Bellonci, P the posterior thalamic nucleus, U the uncinate pretectal nucleus, and X the basal optic nucleus. The large egg-shaped structure is the optic tectum. Reprinted by permission from Scalia and Fite [1974].

Rana esculenta, the horopter passes through the eyes, is circular in cross section, and has a diameter of from 5 to 10 cm in the longitudinal plane [Gaillard and Galand, 1980].

The indirect ipsilateral projection originates in the contralateral tectal hemisphere and is transmitted via the contralateral nucleus isthmi. Neural pathways responsible for this projection have been studied by Fite and Scalia [1976], Glasser and Ingle [1978], Gruberg and Udin [1978], Gruberg and Lettvin [1980], Wang et al. [1981], and Grobstein and Comer [1983]. Some of the efferent fibers from the tectum project to the ipsilateral nucleus isthmi. This nucleus, in turn, sends its fibers to both the ipsilateral and the contralateral tecta. The resulting connections yield the visuotopic mappings shown in Fig. 2.8.

2.3.4 Binocular units in tectum and thalamus

Electrode recordings have been made from binocularly active neuronal units in both tectum and posterior thalamus. The discovery of units that respond to stimulation from either eye shows that tectum and posterior thalamus are not only sites for the reception of binocular stimulation but are also involved in the integration of binocular information.

Since frogs and toads neither exhibit vergence nor have a distinct foveal area, it is unlikely that their binocular depth resolution is based upon finely tuned disparity sensitive units. If two eyes converge to fixate the same small region of visual space on the foveas of their two retinas, then the range of disparities to be considered during binocular matching will be small and centered about zero. This range is known as Panum's fusional area [Panum, 1858].

Without vergence, the range of disparities to be considered must cover not just those within a narrow fusional area but must extend over a large portion of the retinal surface. Electrode recordings within both tectum and thalamus of frogs confirm this supposition. Fite [1969], and later Skarf and Jacobson [1974], recorded from binocular cells in deep tectal layers. These cells have large receptive fields (mean 76°), and tend to be multimodal, responding to vibratory, tactile, and auditory stimulation as well as to visual stimulation. Brown and Marks [1977], recording in thalamus, demonstrated the existence of similar multimodal binocularly sensitive units having even larger receptive fields. Raybourn [1975], and Finch and Collett [1983] reported other, more narrowly tuned binocular units in tectum. Both the tectal and thalamic binocular units exhibit spike frequencies that indicate a summation of contralateral and ipsilateral stimulation, and thus they are responsive over a broad range of disparities. This is in contrast to the excitatory/inhibitory interactions found in mammalian disparity-tuned units [Poggio, 1979].

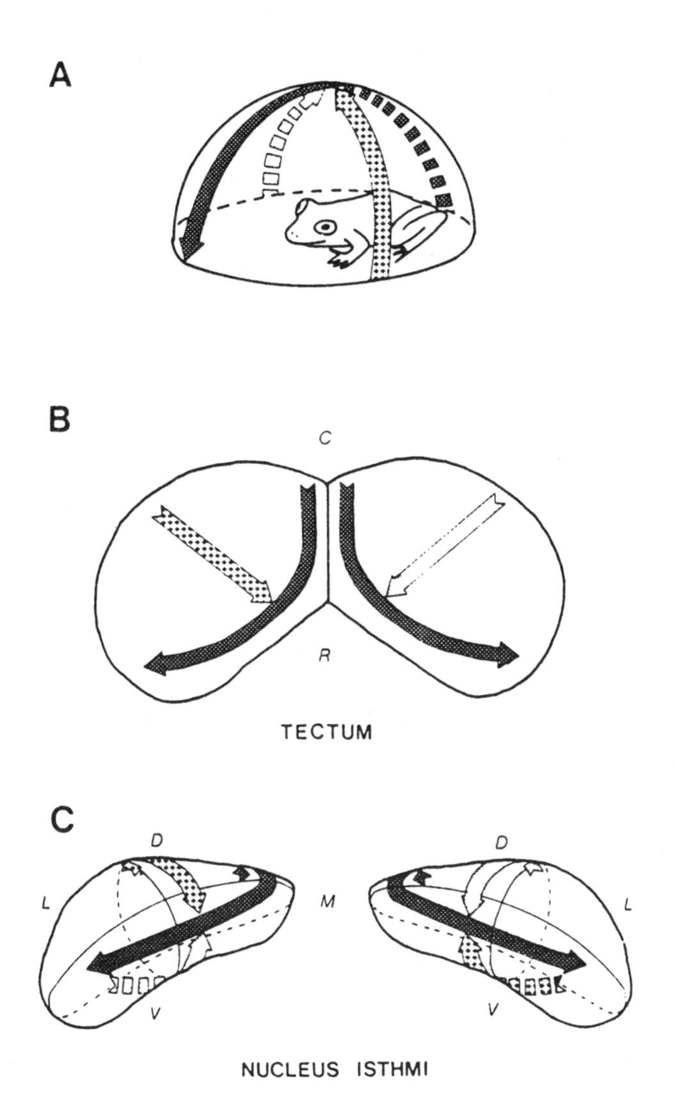

FIGURE 2.8. Retino-Tectal and Isthmio-Tectal Projections
(a) Arrow represents the upper visual field. (b) Dorsal view of tectum (caudal end upward) showing the projection of the visual field formed by terminals of retinal ganglion cells. Arrows correspond to those in (a). (c) Nuclei isthmi represented as shells in a view facing the rostral pole. M, L, D, and V indicate medial, lateral, dorsal and ventral positions. Arrows indicate locations in the contralateral tectum to which cells on the surface of the nuclei project. Reprinted by permission from Gruberg and Udin [1978].

2.3.5 Efferent Pathways from Tectum and Thalamus

Tectal efferent pathways ascend into the thalamus and descend to both the nucleus isthmi and the brainstem motor nuclei. In this study we will be most interested in the descending pathways. Rubinson [1968] used fiber degeneration techniques to trace the anatomy of these pathways in *Rana pipiens*. He found two distinct descending paths. One path extends ipsilaterally to the lateral brainstem tegmentum and the nucleus isthmi. The other path projects contralaterally to the medial brainstem tegmentum and spinal cord. (See also Corvaja and d'Ascanio [1981]).

Two major efferent pathways also project from the thalamo-pretectal region to downstream brain regions. Ingle [1983] used horseradish peroxidase (HRP) staining in bilaterally-tectal-ablated frogs to trace these two pathways from the pretectum into the medulla. He found that an ipsilateral descending route converges with the tectobulbar route through the medulla. A second route crosses the midline at the ventrolateral boundary of the tegmentum and then runs down the ventromedial wall. Later, by HRP backfilling, Ingle determined that a dense posterior medial cell group in pretectum contributes to both the medial (crossed) and lateral (ipsilateral) paths, and that a more lateral pretectal cell group contributes only to the ipsilateral path.

The pathways that descend to the brainstem from both the thalamo-pretectal region and tectum share a certain symmetry. In both cases, there are ipsilateral as well as crossed pathways. Cell layers closest to optic nerve terminals contribute mainly to the ipsilateral pathways. Denser cell layers, farther from the terminals, contribute to both ipsilateral and crossed pathways. Fig. 2.9 (from Ingle [1983]), depicts these pathways. Besides showing the crossed vs. ipsilateral symmetry of the pathways, this figure indicates that the tectal crossed pathway passes the midline at the ventral surface of the tegmentum in the isthmial region. The pretectal pathway crosses the midline caudal to the tectum in the area of the isthmus.

2.4 Functional Analysis of the Major Visuomotor Centers

2.4.1 Retina

In addition to its obvious function as the visual receptor surface of the brain, the retina of frogs and toads plays a role in pattern recognition. Pattern selectivity inherent in the retina was investigated in the frog earlier than in any other species. In their pioneering study, Lettvin et al. [1959] recorded from both the optic nerve and the tectum of frogs, *Rana pipiens*, while moving objects on the inside of a hemisphere that had been aligned

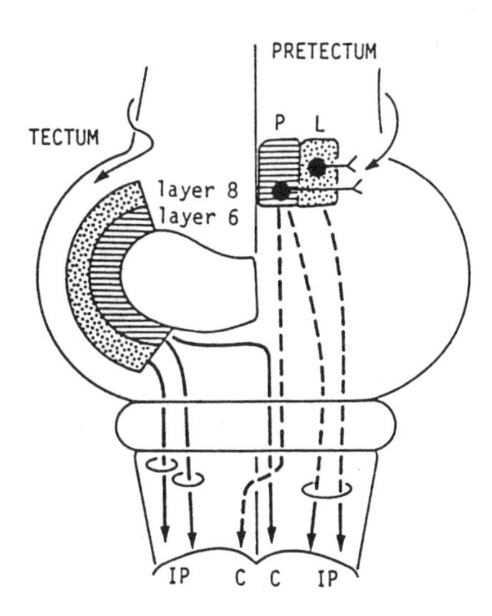

FIGURE 2.9. Tectal and Thalamo-Pretectal Efferent Pathways
Solid lines indicate descending motor pathways from tectum, dashed lines from thalamus-pretectum (TPT). The most superficial tectal efferent layers (8 and 7) project ipsilaterally (IP). Deeper layers (6) contribute to both ipsilateral and crossed (C) pathways. Projections from TPT project ipsilaterally from sites near to optic nerve terminals (L) with more distant somas (P) sending axons along both ipsilateral and crossed paths. Reprinted by permission from Ingle [1983].

with the optical axis of one eye. In this way they were able to identify four distinct classes of retinal ganglion cell. Type 1, *sustained contrast detectors*, have the smallest receptive fields (nominally 2°) and respond vigorously to a light-dark edge that is moved into the receptive field and then held stationary. Type 2, *net convexity detectors*, have larger receptive fields (7°) and respond particularly well to small objects moving with respect to the background. Type 3, *moving-edge detectors*, respond to an edge moving through their 12° receptive fields. Type 4, *net dimming detectors*, have the largest receptive fields (about 15°). Cells of this type respond to large dark objects entering the receptive field or to a rapid reduction in general lighting level. The discovery of these pattern selective units demonstrates that the retina must be considered as a pattern-discriminating device and not merely a light-gathering device.

Because of retinal pattern selectivity, it is unlikely that visual stimulation in the brains of frogs and toads consists of a fine-grained pattern that encodes retinal illumination level. It is more likely that certain visual stimuli are emphasized and other stimuli suppressed so that the total stimulation pattern is rather sparse. Although Fischer [1973] has shown that it is possible to reconstruct a fine-grained image from pattern-selective input in cats, it is a matter of conjecture whether or not frogs and toads utilize such a reconstructive process.

It is an oversimplification to state that the retina is capable of discriminating, for instance, prey from barrier. However, this sort of sensory and functional differentiation is observable in tectum and thalamus.

2.4.2 FUNCTIONS OF THALAMUS AND TECTUM

There is extensive evidence associating the tectum with prey detection and the thalamus with barrier detection. For instance, electrophysiological studies show that tectal visual units all seem to be motion-sensitive [Grüsser and Grüsser-Cornehls, 1976], but there are at least two classes of thalamic units that are sensitive to stationary stimuli. These are the blue-sensitive "on" units reported by Muntz [1962], and stationary edge detectors reported by Ewert [1971], Brown and Marks [1977], and Ingle [1980]. This difference between tectal and thalamic units suggests that these two areas may be responsible for the processing of very different visual stimuli. In particular, they implicate tectum in prey detection and thalamus in barrier detection. Lesion studies confirm this supposition.

Evidence found by Ingle [1977] during an optic nerve regeneration study strongly implicates the tectum in prey recognition. In his experiment, Ingle performed unilateral tectal ablations in frogs and then allowed time for the optic nerve to regenerate to the opposite tectal surface. Frogs "rewired" in this way would respond to prey dummies presented in the visual field contralateral to the ablation as if they were in the opposite field. Their wrong-way turning and snapping was calibrated with the position of the prey in the visual field — the more eccentric the target the greater the amplitude of the misdirected turn.

The behavior of Ingle's "rewired" animals in the presence of barriers strengthens the argument that barrier navigation is not tectally mediated. These animals showed strong dissociation of function between orientation to prey and navigation around barriers. With a barrier placed in the unaffected visual field and a prey in the affected field, these frogs initiated turns around the barrier as if the prey were behind the barrier. With the sides of prey and barrier reversed, the same animals made normal orientation turns toward the prey, ignoring the barrier in the opposite visual field.

In experiments with atectal frogs, Ingle [1982] was again able to demonstrate the division of function between thalamus and tectum. Bilateral tectal ablation abolished both orienting and prey catching responses. However, the same animals were still completely reliable in clearing barriers to avoid noxious tactile stimulation. Their behavior was such as to just clear the barrier edge, whereas intact animals invariably turned away from the threat.

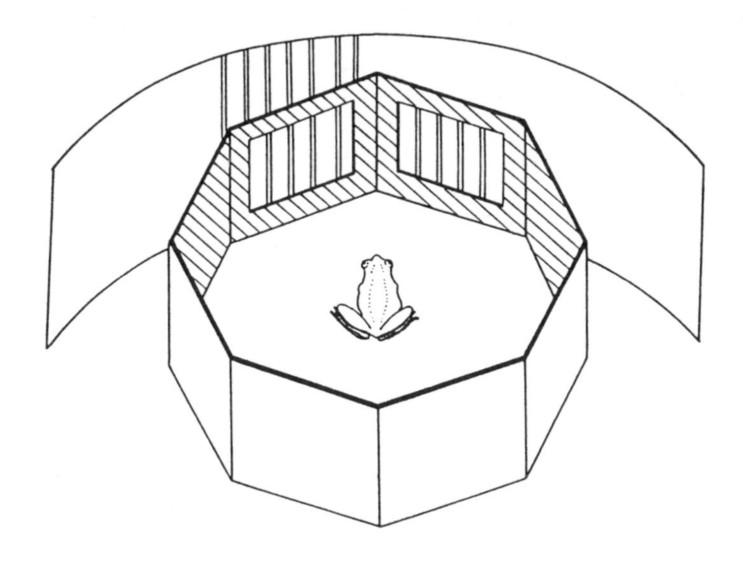

FIGURE 2.10. Depth Vision in Atectal Frogs
Reprinted by permission from Ingle [1983].

2.4.3 Evidence for depth perception in thalamus

Ingle [1982] also showed that his atectal frogs were able to discriminate the depth of barrier surfaces during avoidance behavior. Fig. 2.10 shows the experimental design used to demonstrate this. An opaque barrier placed around the frog's head had one opening, through which could be viewed a series of vertical bars placed on a surface several inches behind the enclosure. Near this opening was a similar configuration of vertical bars, but these bars were placed on the surface of the enclosure. The bars in this second set were calibrated so that their size and spacing would appear to be identical to the bars visible through the opening. During avoidance trials the frogs preferred the real opening as an escape route, rarely orienting toward the false opening. This experiment clearly showed that frogs can discriminate the depth of surfaces without the use of tectum. Since it receives binocular innervation, the thalamus is implicated in this discrimination.

2.4.4 Motor pathways from tectum and thalamus

Studies of the pathways descending from both tectum and thalamus not only confirm the thesis segregating prey detection to tectum and barrier detection to thalamus but also give some indication of the types of motor activity controlled by these two areas. Studies by both Ingle [1983] and Grobstein et al. [1983] showed that the crossed and uncrossed tracts descending from both the thalamo-pretectal area and the tectum perform distinctly different functions.

The crossed pathways from tectum have been implicated in the orienting response. Ingle [1983] investigated the effects of severing the crossed descending pathways from tectum by lesioning the ansulate commissure at the ventral surface of the tegmentum. Frogs prepared in this way were highly motivated to strike at prey but could not orient toward their prey. Snaps were always straight ahead. For more distant prey they hopped forward, again without the usual orienting response. However, for prey positioned above the head they were able to make the appropriate vertical movements.

If the crossed descending tract from the pretectal area is severed, frogs lose their ability to sidestep around barriers. Ingle transected the crossed fibers from pretectum by splitting the isthmus. These frogs retained their ability to turn toward prey. However, they were unable to turn or sidestep to negotiate barriers.

Frogs with severed crossed pretecto-fugal pathways show a motor but not a sensory deficit. For example, although a tail pinch would send them crashing into a barrier if it crossed the midline by more than 15°, the same animals were able to round a barrier to catch prey if the barrier edge was on the midline, so that a sidestepping movement was not required. Also, the lesioned animals would consistently suppress strikes toward prey located behind a frontal barrier.

The evidence from his lesion experiments led Ingle to make the following functional description of the crossed and uncrossed tecto- and pretecto-fugal pathways. First, the uncrossed (ipsilateral) pathways are involved in symmetric motor behavior, i.e. snapping or walking straight ahead or moving the head and body vertically. Thus, the ipsilateral tectal pathways govern strike versus no-strike activity and the ipsilateral pretectal pathways govern avoidance versus nonavoidance of a barrier surface. The crossed pathways, on the other hand, govern one-sided or asymmetrical movements. For tectum they direct turning or orienting movements toward prey, and for pretectum they govern sidestepping or turning around barriers.

2.4.5 Depth perception dissociated from orientation

Ingle's experiments with split-tegmentum frogs also showed that these animals retain their ability both to estimate depth and to use this depth information to regulate their behavior. Frogs will snap at a prey object if it is within a frontal "snap zone," but prefer to hop toward a prey if it is located beyond this zone [Ingle, 1982]. Ingle's split tegmentum frogs would not turn toward prey; all snaps were straight ahead. However, even though they could not aim correctly these frogs would still snap when a prey object was located anywhere within the snap zone and hop forward when a prey object was located outside of the snap zone.

2.4.6 Depth Maps, Motor activity and tectal retinotopy

Although tectum appears to play a key role in determining depth for prey catching, recent experiments by Collett and Udin [1983] indicate that it is unlikely that a tectal depth map is used to supply this information. They were able to demonstrate that toads with large electrolytic lesions to nucleus isthmi, and thus deprived of binocular input to tectum, were still able to use binocular cues to judge prey depth. Collett and Udin's lesioned animals were as accurate as controls in snapping at a single prey object, and when base-out prisms were placed in front of their eyes the lesioned toads made the undershooting errors predicted for binocular depth estimation.

Although early evidence indicated that the tectum directed motor output and that it did this in a retinotopically organized way, more recent results appear to refute this claim. By means of electrode stimulation in tectum Ewert [1971] was able to elicit typical prey-catching responses that were consistently directed at the spatial location predicted from the organization of the retinotectal map. However, Grobstein et al. [1983] found that the motor pattern generators underlying the prey-catching behavior of frogs do not reside in tectum. They found that, although tectal lesions abolish visually elicited orientation responses, orientation to tactile stimuli is preserved. Conversely, lesions to *lateral torus semicirculus* abolish tactilely induced orientation while sparing visually induced orientation. Thus they concluded that the motor pattern generators must reside in a brain region that receives input from both tectum and torus.

In the same series of experiments, Grobstein et al. also found that, although visual input to tectum is retinotopically organized, this retinotopy is not preserved in the tectal output. Small lesions that cut across part of the tecto-motor neuraxis did not produce the expected orientation scotomas. Instead, frogs with these lesions tended to undershoot the stimulus position when orienting to a stimulus contralateral to the lesion. However, even though they lost accuracy, the lesioned animals continued to be able to initiate orientation turns to all angular directions. It was not until the neuraxis was completely severed that orienting responses contralateral to the lesion were abolished. If a retinotopically organized map from tectum to motor output did exist, the lesions to the neuraxis would have had distinct local effects throughout the affected hemifield. Instead, the effect appeared to be distributed, with the number of intact fibers governing accuracy over the entire region.

The idea of a retinotopically organized map from tectum to motor output was even more directly challenged by a second experiment [Grobstein et al., 1983]. In this experiment, both a unilateral tectal lobe removal and a hemisection of the neuraxis were performed. Thus when the tectal removal and hemisection were done on the right side, tectal activity from the left eye's visual field was destroyed and the crossed descending pathway from

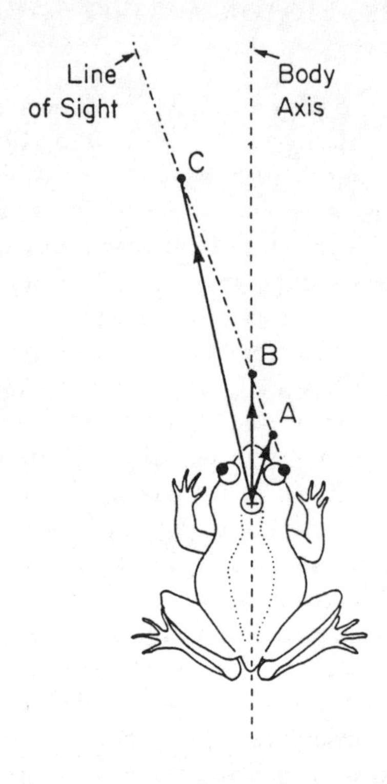

FIGURE 2.11. Orientation Must be Spatially Directed
Orientation responses (arrows) to ojbects at points A, B, and C. Redrawn by permission from Grobstein et al. [1983].

the left tectal lobe was also severed. The only descending tectal pathway remaining intact was the uncrossed ipsilateral pathway on the left side (see Fig. 2.9). As would be expected, these animals could not make visually elicited turns to the side ipsilateral to the lesions. However, they were able to make turns to the side contralateral to the lesions but only in response to an ipsilaterally located stimulus. The amplitude of the wrong-way turn could be increased by placing the stimulus at a more eccentric position in the opposite visual field. The fact that these unitectal animals could elicit turns into the visual field covered only by the ablated tectal lobe makes it highly unlikely that each tectal locus can be associated with a turn of a particular direction.

The argument that orientation turning cannot be topographically organized can also be made from considerations of visuomotor geometry. Since the body-turning axis is offset from the origin of an eye-centered coordinate system, various orientation responses are appropriate for a single retinal location. Fig. 2.11 from Grobstein et al. [1983] demonstrates this point. Objects A, B, and C all stimulate the same retinal locus and thus the same tectal position. However, with the body-turning axis placed on the midline but behind the plane of the pupils, A results in a right turn, B in no turn, and C in a left turn. This ambiguity in relating turn direction to tectal topography can only be resolved by considering target depth as well as angular position.

2.4.7 Summary of the Frog/Toad Visual System

In our discussion of the visual system of frogs and toads we have made the following points: (1) frogs and toads make use of depth vision to locate both prey and barriers; (2) they utilize both binocular and monocular depth cues; (3) monocular depth cues are a result of lens accommodation; (4) there is a substantial region of binocular overlap in the visual field; (5) there are retinotopically organized visual maps in both thalamus and tectum; (6) the visual maps cannot be assumed to represent local illumination level, since the visual input to the brain is feature-encoded; (7) both tectum and thalamus receive binocular visual input and both have binocularly activated neuronal units; (8) tectal binocular input is via an indirect relay in nucleus isthmi but thalamic binocular input is direct; (9) tectum is implicated in the recognition and localization of prey, and thalamus is implicated in the detection and localization of barrier surfaces; (10) it is unlikely that a tectal depth map is used to direct prey catching; (11) retinotopy may not be preserved in the efferent pathways from tectum; and (12) orientation turning depends on the depth of a target as well as on the position to which it projects in tectum.

2.5 Modeling Assumptions

Four assumptions made in the depth models discussed earlier are inconsistent with the data on the visual system of frogs and toads. Three of these assumptions have to do with the nature of the images available to the depth resolving system. These are assumptions of vergence, static images, and dense images. The fourth assumption is more fundamental. It is the assumption that the model's purpose is to produce a complete depth-mapping of the image surface.

All of the depth models are based on the assumption that the image pair to be processed has been obtained from an imaging system with the capability of vergence. Due to vergence, the disparities in the image will be small and centered about zero. Thus the matching algorithm can be constrained to consider only those matches that lie within a narrow disparity range and the smaller disparities can be weighted more heavily than the larger ones. Since frogs and toads do not demonstrate vergence, this is an inappropriate assumption for a model of their process of depth perception.

Since the models operate on a static image pair, the only available depth cues are those that can be obtained directly from these images. Any active process that a functioning animal might utilize to explore its visual surroundings cannot be considered. This restriction explicitly rules out treatment of depth cues that could come from motion parallax due to eye or head movements, change in image size during approach, or change in image focus due to lens accommodation. It is known that frogs and toads use lens accommodation as part of their depth perception process. Thus, at least this active process must be incorporated in the model.

The models also depend upon there being intensity variations between neighboring picture elements across most of the image surface. For example, most of the models were tested using either random-dot stereograms or digitized photographs. It is not clear that any of the proposed mechanisms would be appropriate for producing a depth map from the more sparse feature-encoded visuotopic maps available to frogs and toads.

Finally, all of the previous depth models were designed to build a visuotopically organized depth map. None of these models consider the end use of the depth information that they produce. The intent in developing these models was to provide a general solution to the binocular depth resolution problem and not to design an algorithm that could be integrated into the activity pattern of a specific animal. In our study of depth perception we wish to define algorithms that, even at the expense of generality, are consistent with both the known behavior and the neural substrates of frogs and toads.

We have defined some of the constraints that guided our efforts at modeling depth perception in frogs and toads. In the remaining chapters we develop two quite different models guided, more or less, by these constraints. Finally, we contrast the models and show how the actual depth perception process might make use of elements of both models. The first of these models does make use of the concept of a general depth map. It falls naturally in the family of models discussed at the beginning of this chapter. It uses cooperative computation to build its depth maps using both binocular disparity matching and lens accommodation as depth cues. An important feature of this model is that it can work both with and without binocular depth cues. The second model avoids the use of a depth map and instead provides the spatial location of a single point in the visual field. This model is much more specific about the interaction of binocular and lens accommodation mechanisms. It represents this interaction as part of an action-oriented feedback control loop which is specialized to prey capture.

2.6 Conclusions

Our comparison of the anatomy, physiology, and behavior of frogs and toads with a series of recent models of binocular depth perception has shown that none of these models can adequately represent the depth perception process in frogs and toads. We have noted that a successful model (1) must not depend upon vergence; (2) must be able to account for the use of active processes, particularly lens accommodation; (3) must be able to operate on sparse images; and (4) must not depend upon a tectal depth map. Our discussion of the visual system of these animals argues that (1) the process used to estimate the depth of barrier objects might be very different from that used to estimate the depth of prey, (2) prey-depth discrimination is tectally mediated whereas barrier depth is determined in thalamus, and (3) both binocular and monocular (from lens accommodation) depth cues are used in prey-depth determination.

3

Monocular and Binocular Cooperation

ABSTRACT In this chapter we introduce a model of depth perception that explores one way in which lens accommodation cues can be used to help disambiguate the correspondence problem of stereopsis. Although the model was originally developed to explain how frogs and toads accomplish binocular depth perception without the use of cues from vergence, it is also of general interest as an extension to the class of cooperative stereo models. It consists of two cooperative depth discrimination processes, each acting to build a depth map, with one process receiving monocular depth cues based on accommodation and the other receiving binocular depth cues based on disparity. Mutual excitatory connections between the maps allow the model to converge to a single solution where accuracy is governed by binocularity. The model will also function in a purely monocular mode if binocular input is removed. Neurophysiological data on the visual system of frogs and toads are used to constrain choices made in constructing the model, and results obtained from simulation runs are compared with data from behavioral experiments.

3.1 Design of the Model

The problem of the computation of a depth map from the binocular correspondence of disparate retinal images is complicated by the potential for ambiguity in the process of pairing local features. Fig. 3.1 shows the simplest example of this ambiguity. The spatial points A and B cannot be distinguished from the "ghost" targets (C and D) by simply considering the positions of stimulation on the two retinal surfaces. Because of this ambiguity, a system for stereoscopic depth perception must have sources of disambiguating information. In Chapter 2 we showed that most sources for disambiguating cues proposed in previous models of stereoscopic depth perception come either explicitly or implicitly from the assumption of binocular convergence.

In the development of our model we wished both to adhere to the general methodology employed in earlier cooperative stereo models and to address the unique features of the depth resolving system of frogs and toads. Our goals were (1) to eliminate dependency on vergence by using monocular cues from lens accommodation to help disambiguate binocular cues from image matching; and (2) to define a mechanism that can function when

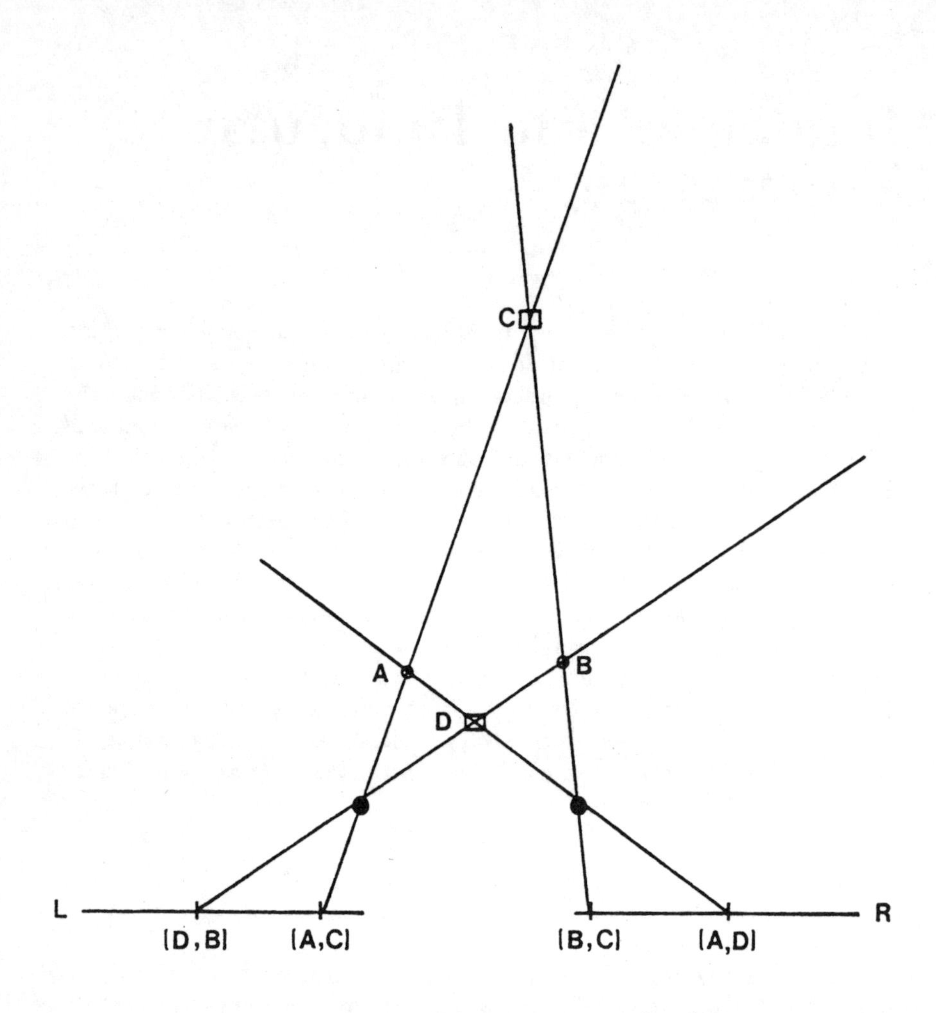

FIGURE 3.1. Ambiguity in Binocular Image Matching
L and R represent left and right retinal surfaces, filled circles are pupils, and lines indicate lines of sight.

only monocular depth cues are available, even though binocular depth cues dominate.

3.1.1 Structure

In place of a single depth-resolving process utilizing only binocular depth cues, our model employs a pair of interacting processes, each with its own source of depth cues. One process receives monocular depth cues from lens accommodation, and the other receives binocular depth cues based on disparity matching. The two processes cooperate to augment each other.

Fig. 3.2 depicts the overall connectivity of the model. Layer A represents an inference system that provides monocular depth cues from lens accommodation, and layer D represents an inference system that provides binocular cues from disparity matching. Layers M and S represent spatially organized fields over which two depth-mapping processes operate. Arrows from layers A and D to these fields indicate that field M receives only monocular depth cues, and field S receives only binocular cues. The ovals and arrows between fields M and S indicate mutual excitatory interconnections that map each local region (oval) of one field onto the corresponding local region of the other field. By means of these interconnections, points of high excitation in one field provide additional excitation to corresponding points in the other field. Thus there is a high degree of synergy wherever similar points are externally excited in both fields. Competition among depth estimates within each field assures that points excited in only one field will have little chance to sustain this excitation when there are other points receiving stimulation in both fields.

In this system, monocular cues from lens accommodation need not be used to estimate depth directly. Binocular depth cues are ambiguous simply because two or more cues at the same retinal position can have identical weights. Accommodative cues may be quite weak, compared with the binocular cues, and still provide enough of a bias to resolve the binocular ambiguity. If the model is tuned in this way, accommodative cues need not be acute, or even very accurate, for the scheme to work properly. Since accommodative cues simply serve to augment corresponding binocular cues, the accuracy of the system will be dominated by the accuracy of the binocular estimates. However, since the accommodative cues are not ambiguous, they can be used to estimate depth, even in isolation from binocular cues.

Fig. 3.2 shows the efferent pathways from the model emanating from the monocularly driven map. This is meant to indicate that the monocularly driven map should be capable of directing motor coordination whether or not binocular information is available, even though binocular information dominates when it is available. In actual practice it makes little difference which of the two maps provides the efferents since, when the model is tuned for good performance, the two maps tend towards nearly identical equilibrium states.

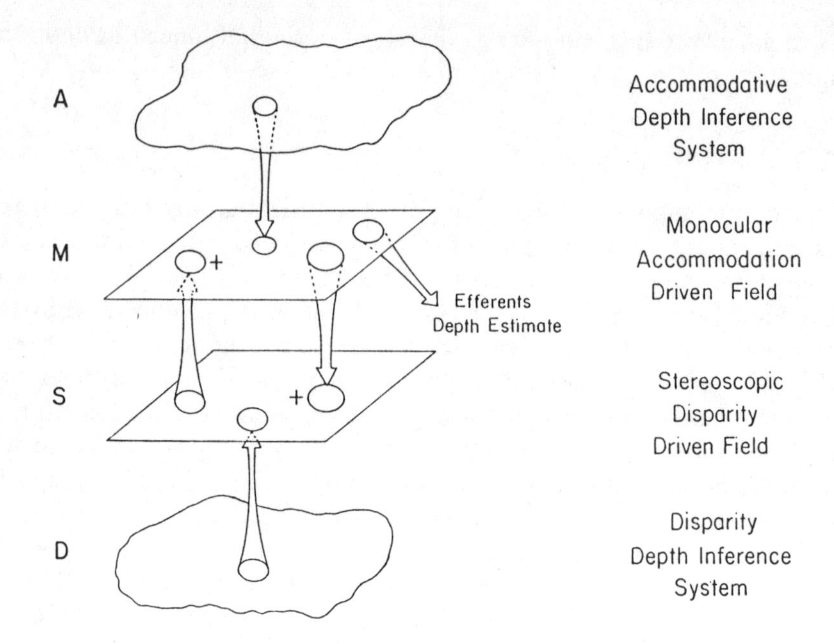

FIGURE 3.2. Connectivity of the Complete Cooperative Depth System

3.1.2 CONSTRAINTS

In designing our model of depth perception we wished to adhere to the underlying principles of previous cooperative depth models but at the same time to incorporate the additional constraints derived from the visual system of frogs and toads. The prime constraints of most cooperative stereo models are (1) uniqueness: for each retinal position only one depth may be assigned; and (2) continuity: disparity (or depth) varies only gradually or not at all over most of the image [Dev, 1975; Marr and Poggio, 1976]. The constraints derived from the properties of the frog/toad visual system are (3) the ability to function both with and without binocular depth cues; (4) the lack of binocular depth cues from vergence, together with the availability of monocular cues from lens accommodation; (5) sparsity of visual input due to the presence of pattern-sensitive mechanisms in the retina; and (6) the separation of visual input into distinct prey and barrier channels that project to two different brain regions.

All of the above constraints are mutually consistent, with the exception of continuity (2) and sparsity (5). Continuity implies the availability of dense images so that regions of depth discontinuity occur only at the edges of objects. Sparsity, however, implies that the image is mostly void, with the exception of a few areas of select stimulation. This conflict was resolved by relaxing the continuity constraint to require only that most neighboring *stimulated* points share similar disparities.

Constraint (6), that the visual input is broken into separate prey and barrier channels, has three possible implications for depth perception in frogs and toads. The first is that the process of depth perception might take place before the functional differentiation occurs. This is deemed unlikely since atectal frogs are able to judge the depth of barrier surfaces during escape, but the same animals completely ignore prey objects [Ingle, 1982]. The other two possibilities are that identical depth perception processes are replicated in two different brain regions or that the process for estimating depth of prey is different from the process for estimating depth of barriers. For the remainder of this chapter, we proceed as if two identical processes are used. This is done for theoretical reasons alone, and does not imply a belief that the two process are, in fact, identical. Treating both processes identically simply serves as a means for obtaining data that allow us to compare the functioning of the model on two very different kinds of visual input. In Chapter 6 we explore this matter further.

3.1.3 Coordinate systems

Although previous stereoscopic depth perception models have all employed a single cyclopean coordinate system, we chose eye-centered coordinate systems for the expression of our model. These coordinate systems plot angular disparity versus retinal angle separately for each eye. The choice of two eye-centered coordinate systems was made in order to provide a representation of visual space appropriate for treating both binocular and monocular depth cues. Since both the cyclopean and the eye-centered coordinate systems are based on disparity, they are equally appropriate for representing disparity-cued depth measurements. However, depth cues based on accommodation are naturally eye-centered.

The problem was restricted to the consideration of one-dimensional retinas and a two-dimensional world, allowing these coordinate systems to be constructed as shown in Fig. 3.3a. For the right-eyed system, the horizontal axis q is defined by angular position on the right retina R, so that a point on the right retina maps to a vertical line. The vertical axis d measures angular displacement of the image on the left retina L from the image appearing at the indicated position on the right retina, so that a point on the left retina maps to a diagonal. This displacement, or disparity, is taken as being positive when the portion of the image being considered on the left retina is to the right of that on the right retina. The left-eyed system is configured similarly but with the horizontal axis measuring position on the left retina.

Fig. 3.3b illustrates how the right-eyed system would appear if it were overlaid on an external Cartesian system (see Appendix B). The spacing of the equi-disparity arcs illustrates the decrease of depth acuity with increasing distance from the animal. The properties of optical depth resolving systems are reflected in the compression of the representation near to the eye, allowing fine depth resolution only for nearby objects.

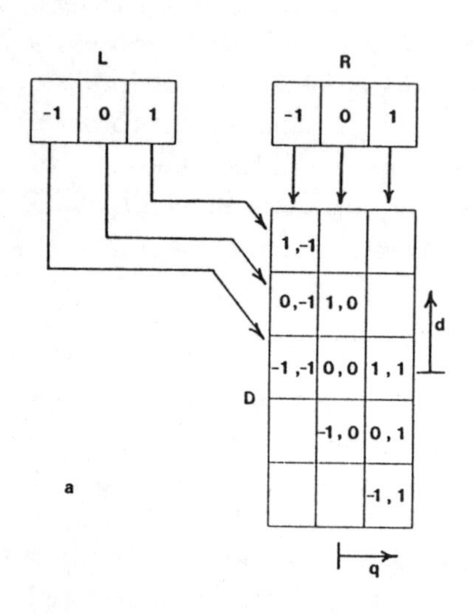

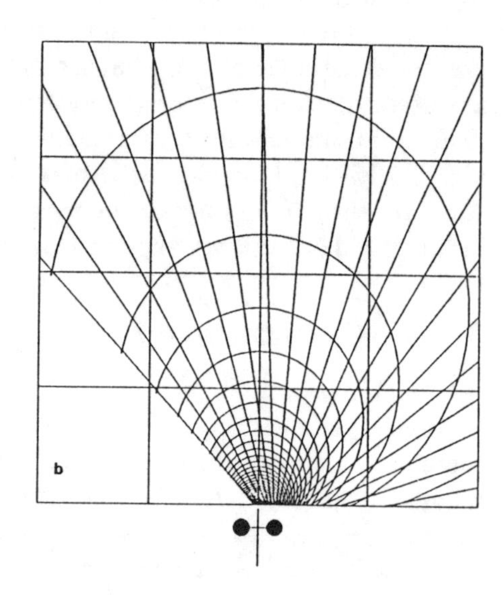

FIGURE 3.3. The Coordinate Systems of the Model
(a) Construction of binocular depth estimates in retinal angle vs. disparity coordinates for the right-eye centered system. Array D represents a continuous plane and number pairs within the cells of D indicate which points on the left and right retinas are represented by the cell of D in which they appear. (b) Projection of the right-eyed coordinate system back onto an external Cartesian grid. Radial lines are at equal angular increments; arcs are lines of constant disparity spaced at equal increments of disparity. The object below the grid area represents a frog or toad with eye positions indicated by the large disks.

3.1.4 COMPUTATIONAL SCHEME

A general cooperative/competitive scheme, devised by Amari and Arbib [1977] and based upon earlier work by Didday [1970] and Dev [1975], was used as a substrate for the model. Amari and Arbib consider situations where the input is a two-dimensional array, and the problem is to select a single distinguished element for each column of this array. Their solution consists of a homogeneous two-dimensional excitatory neural field interacting with a one-dimensional inhibitory neuron pool. Since this scheme selects a single element for each column of the input, it incorporates the uniqueness constraint (1). The continuity constraint (2) is realized by mutual excitatory connections between neighboring columns of the field.

Fig. 3.4a shows how the Amari/Arbib scheme handles the depth perception problem for the case of the projection of a two-dimensional scene onto one-dimensional retinas. The excitatory field is defined in the retinal angle q versus disparity d coordinate system. The input to the field is a map that assigns a set of depth-likelihoods to each retinal position, and that is determined by a single depth-inference system. Regions of the field that are close in both retinal position and disparity are mutually excitatory, with the local spread of excitation fairly broad along the retinal position axis and quite narrow along the depth axis. Thus, similar depth estimates at nearby retinal positions cooperate to reinforce each other. The portion of the inhibitory pool assigned to a particular retinal position receives an excitatory signal that is modulated by the sum of all excitation along a narrow band centered about that retinal position in the field. The inhibitory pool, in turn, sends inhibitory feedback to all elements at that retinal position. Thus multiple depth estimates at the same retinal position must compete with each other in order to sustain field excitation. If parameters are properly chosen, the field will reach an equilibrium state where excitation is maintained at only one depth region for each retinal position. Fig. 3.4b shows a cutaway view of the net effect on the field due to both the excitatory and inhibitory connections when a single point source of excitatory input is provided. The inhibitory effect on points at the same retinal position but differing disparities, and the facilitatory effect on neighboring points of similar disparities, are apparent in this figure.

The Amari/Arbib model can be described by the pair of continuous nonlinear integro-differential equations

$$
\begin{aligned}
\mathcal{T}_m \dot{M}(q,d,t) &= -M(q,d,t) + \iint w_m(q-\zeta, d-\eta) f[M(\zeta,\eta,t)]\,d\zeta\, d\eta \\
&\quad -K_m g[U(q,t)] + K_a A(q,d,t), \\
\mathcal{T}_u \dot{U}(q,t) &= -U(q,t) + \int w_u(q-\zeta) g[U(\zeta,t)]\,d\zeta \qquad (3.1)\\
&\quad +K_u \int f[M(q,\eta,t)]\,d\eta.
\end{aligned}
$$

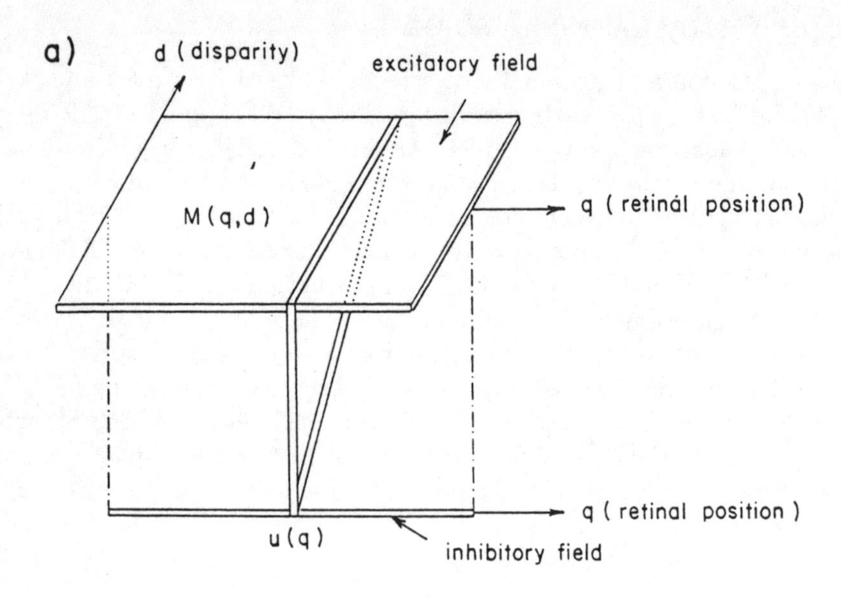

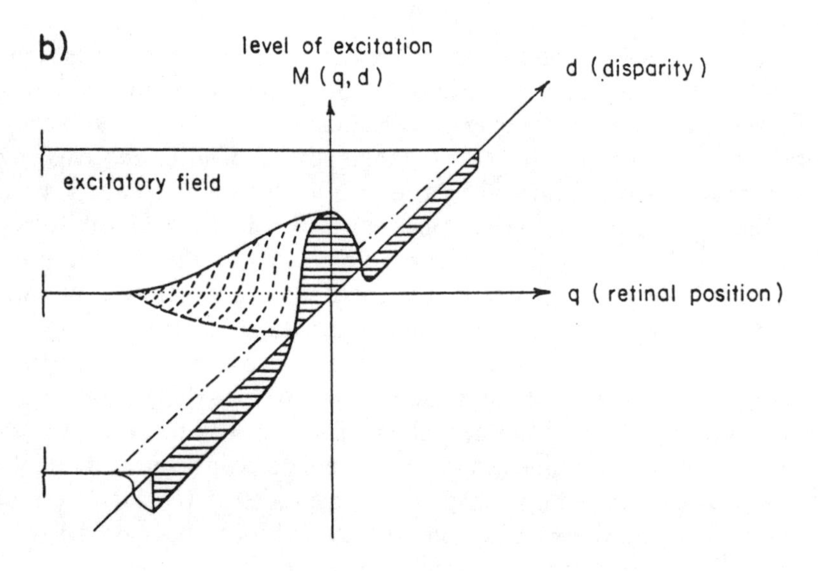

FIGURE 3.4. The Amari/Arbib Model
(a) Structure of the model for the solution of the depth-mapping problem for one-dimensional retinas and one source of depth cues. Redrawn by permission from Amari and Arbib [1977]. (b) Cutaway view showing the net spread of excitation and inhibition around a point-source of external stimulation in the excitatory field.

In these equations we designate the depth inference system by a two-dimensional layer A, the excitatory field by a two-dimensional layer M, and the inhibitory pool by a one-dimensional layer U. The spatial coordinates of layers A and M are retinal angle q and disparity d. The spatial coordinate of layer U is retinal angle q. The point potentials $M(q,d,t)$ and $U(q,t)$ in layers M and U depend on both time t and the spatial coordinates. Internal potential in M is converted to an external potential (or firing rate) by a saturation-threshold function f. Each point in layer M receives excitatory stimulation from the depth inference system A through the gain K_a, and from neighboring points through the spread function w_m. Similarly, the point potential in the inhibitory layer at time t is $U(q,t)$, internal potential is converted to external potential by a simple threshold function g, and mutual excitation between points in U is described by the spread function w_u. Excitatory input to U at retinal position q comes through the gain K_u and is the integral along the disparity coordinate of all excitation at position q in the excitatory field. In turn, layer M receives an inhibitory signal from the inhibitory pool through the gain K_m. The rates of change of potential in layers M and U are governed by the time constants $\mathcal{T}_m$ and $\mathcal{T}_u$.

Our model extends the Amari/Arbib scheme, denoted by eqs. (3.1), by employing two such processes, one receiving monocular depth cues from lens accommodation and the other receiving binocular depth cues from disparity matching. The two processes interact to reinforce similar depth estimates by means of cross-coupling pathways between their excitatory fields. The full depth perception model of Fig. 3.2 is represented analytically by the following four integro-differential equations:

$$\begin{aligned}
\mathcal{T}_m\dot{M}(q,d,t) &= -M(q,d,t) + \iint w_m(q-\zeta, d-\eta) f[M(\zeta,\eta,t)]d\zeta d\eta \\
&\quad + K_{sm} f[S(q,d,t)] - K_m g[U(q,t)] + K_a A(q,d,t), \\
\mathcal{T}_u\dot{U}(q,t) &= -U(q,t) + \int w_u(q-\zeta) g[U(\zeta,t)]d\zeta \\
&\quad + K_u \int f[M(q,\eta,t)]d\eta, \\
\mathcal{T}_s\dot{S}(q,d,t) &= -S(q,d,t) + \iint w_s(q-\zeta, d-\eta) f[S(\zeta,\eta,t)]d\zeta d\eta \\
&\quad + K_{ms} f[M(q,d,t)] - K_s g[V(q,t)] + K_d D(q,d,t), \\
\mathcal{T}_v\dot{V}(q,t) &= -V(q,t) + \int w_v(q-\zeta) g[V(\zeta,t)]d\zeta \\
&\quad + K_v \int f[S(q,\eta,t)]d\eta.
\end{aligned} \tag{3.2}$$

In these equations, A represents the accommodative depth inference system, D the disparity inference system, M the monocularly driven excitatory field, and S the stereoptically driven field. The corresponding inhibitory pools are represented by U and V. Excitatory cross-coupling between the monocularly and binocularly driven processes is represented by additional terms providing stimulation from each point in layer S to the spatially corresponding point of layer M through gain K_{sm}, and from M to S through gain K_{ms}.

Eqs. (3.2) provide a complete analytical description of the model. For a more complete explanation of the basis for this representation, refer to Appendix A.

3.2 Methods

3.2.1 Computer simulation

Computer simulation was used to do the extensive testing and experimentation required for proper evaluation of the model. The equations representing the model were translated into an algorithm, coded in Pascal, and solved numerically on a VAX – 11/780. The simulation algorithm was fully implemented on the computer for both eyes but for only the left side of the brain. It is imbedded in a software system that provides an interactive graphics interface for constructing two-dimensional test scenes, tuning the simulation's parameters, and producing various displays of the simulation's state.

In the simulation system, simple optical equations are used to project a test scene onto two simulated semicircular one-dimensional retinas and to infer the initial accommodative and disparity depth cues. Depth cues from disparity are calculated directly by shifting the images on the two retinas with respect to each other and comparing corresponding points. Consistent with constraint (3), that vergence cues should not be employed, relative shifts of up to 50% of the binocular visual field are used. Accommodation cues are approximated using a Gaussian function centered about the correct depth of a point of stimulation, to generate a set of estimates that is broadly spread in the depth (disparity) direction. Details of the optical equations are provided in Appendix B.

Eqs. (3.2) were simplified for the computer simulation. The simulation represents the two-dimensional layers M, S, A, and D as two-dimensional arrays and the inhibitory layers U and V as vectors. We made the assumption that the excitatory spread in the disparity direction of the two-dimensional layers would be small compared with both the spread in the retinal position direction and the coarseness of the discrete arrays. By this assumption, the double convolution integrals of eqs. (3.2) were reduced to single integrals, and the spread functions w_m and w_s were reduced to functions of only distance along the retinal position axis. Similarly, we assumed that the excitatory spread in the inhibitory pools would be small compared with both the spread in the excitatory layers and the coarseness of the discretization. By this assumption, the spatial convolution integrals were removed from the equations describing the inhibitory layers. After these simplifications are made, eqs. (3.2) are reduced to

$$
\begin{aligned}
\mathcal{T}_m \dot{M}(q,d,t) &= -M(q,d,t) + \int w_m(q-\zeta) f[M(\zeta,d,t)]d\zeta \\
&\quad + K_{sm} f[S(q,d,t)] - K_m g[U(q,t)] + K_a A(q,d,t), \\
\mathcal{T}_u \dot{U}(q,t) &= -U(q,t) + K_u \int f[M(q,\eta,t)]d\eta, \\
\mathcal{T}_s \dot{S}(q,d,t) &= -S(q,d,t) + \int w_s(q-\zeta) f[S(\zeta,d,t)]d\zeta \\
&\quad + K_{ms} f[M(q,d,t)] - K_s g[V(q,t)] + K_d D(q,d,t), \\
\mathcal{T}_v \dot{V}(q,t) &= -V(q,t) + K_v \int f[S(q,\eta,t)]d\eta.
\end{aligned}
\qquad (3.3)
$$

Eqs. (3.3) provide a complete description of the continuous model as it was approximated by the simulation program. However, we wanted to test the model separately against both prey-like and barrier-like visual data in order to emulate the presence of these two channels in the visual system. Thus the visual input to the simulation was broken up into separate "bug" and "barrier" channels, with the "bug" information passed to one complete version of the simulation and the "barrier" information passed to a second version. The two simulations were then run in parallel.

3.2.2 Visual input

A standard test scene was used for most of the experiments with the computer simulation. The configuration of this scene is shown in top view in Fig. 3.5. The simulated animal is directly facing and about 20 cm from the center of a paling fence that is oriented perpendicular to its body axis. Prey objects are located at two different positions behind the fence. This scene was chosen as a composite of typical configurations used in behavioral experiments [Collett, 1982; Collett and Udin, 1983]. The simulation considers only the simplified case of one-dimensional retinas. Retinal images are formed by taking a horizontal slice through the scene.

When processed through the accommodation and disparity-matching depth-inference mechanisms, the standard test scene of Fig. 3.5 produces the input planes shown in Figs. 3.6a - d. These planes are depicted in the retinal-angle q vs. disparity d coordinate system of the model. Surface A shows the extent of the spread of the depth estimates provided by the simulation of accommodation. Surface D shows the ambiguity, due to spurious

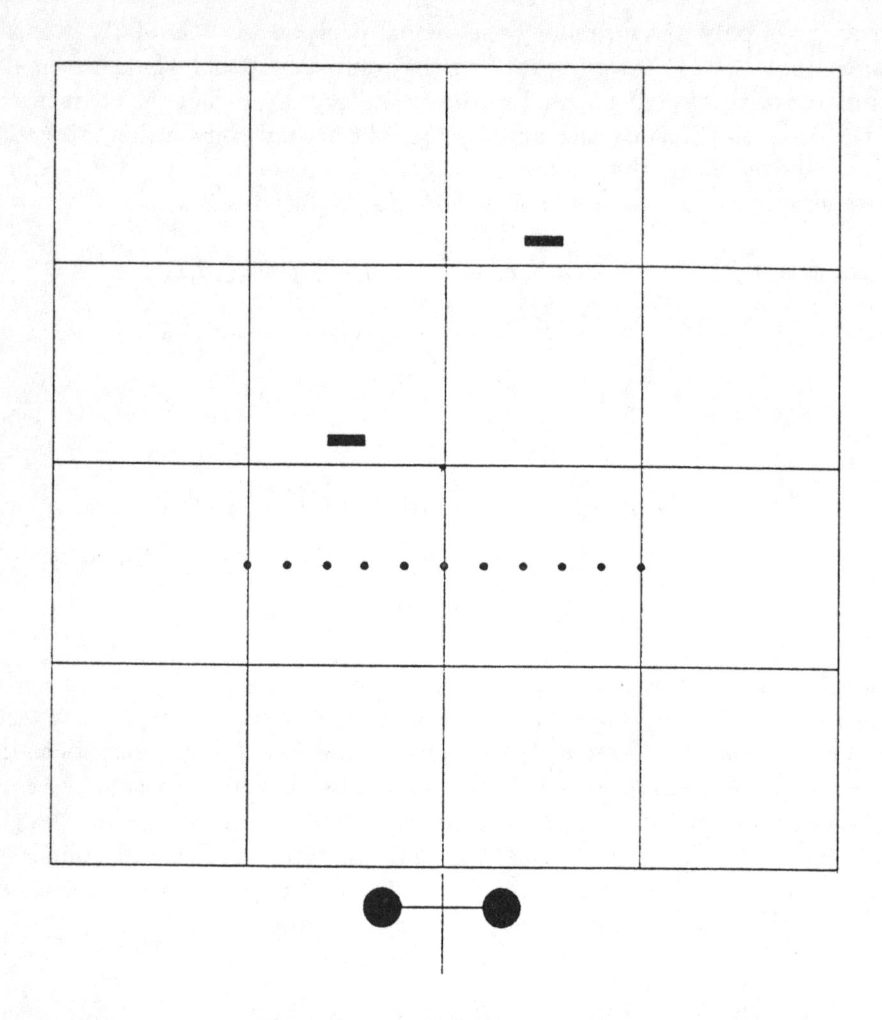

FIGURE 3.5. Standard Test Scene
Each grid element represents a 10 cm by 10 cm area. The simulated frog or toad is indicated by the icon below the grid area; circles indicate eye position. Rectangles represent prey and dots represent a fence interposed between animal and prey.

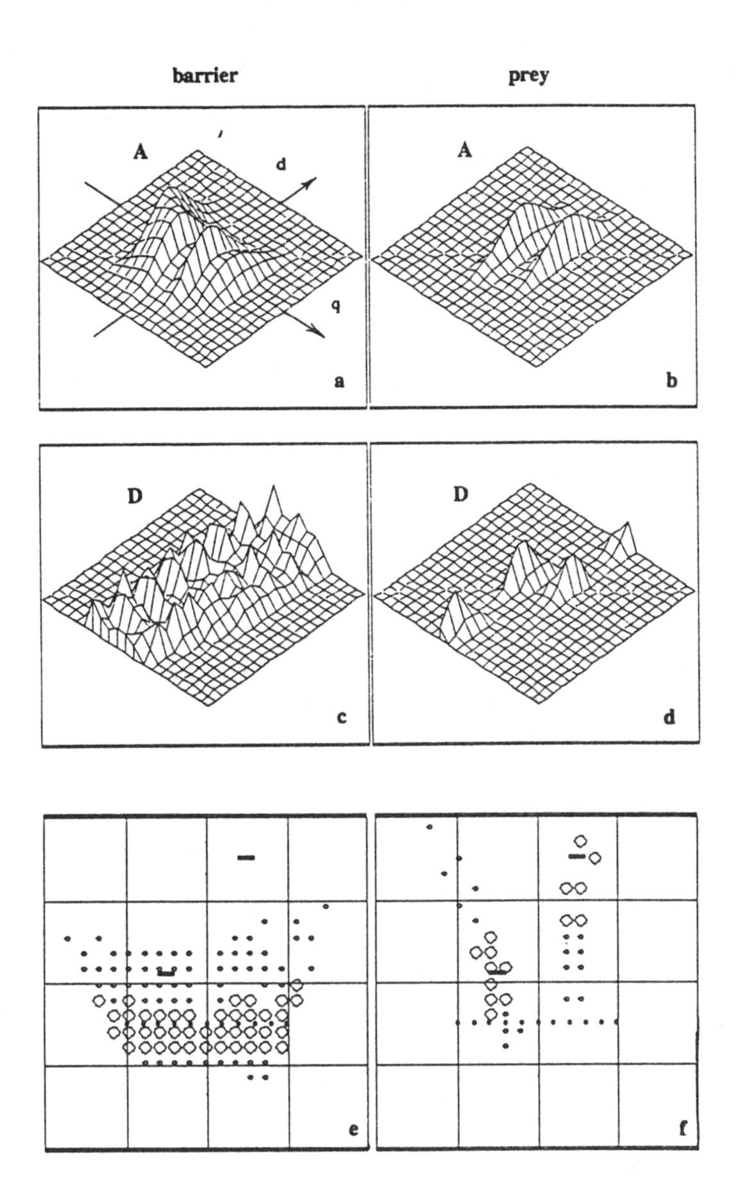

FIGURE 3.6. Monocular and Binocular Inputs — Standard Scene
Initial depth estimates inferred from the standard test scene. (a) - (d) Left-hand figures represent the barrier process and right-hand figures the prey process. Surface A represents the accommodative (monocular), and surface D the disparity (binocular), depth inference systems. (e) and (f) show monocular estimates displayed in the original spatial coordinate system and superimposed upon the test scene.

matching, of estimates provided by disparity. In these figures the height of the curve above the grid-like plane indicates the estimated likelihood that the particular disparity is the correct one for the corresponding retinal position. Figs. 17e - f show how the monocular estimates would appear if transformed back into the external Cartesian coordinate system and overlaid upon the original scene. Relative estimate strength for a particular portion of the visual field is encoded by the size of the small "oval" at that position in the field. These figures give only a coarse visual feel for the extent of uncertainty in the input but provide a means for comparison with the relatively fine discrimination achieved by the model in the studies described below.

3.3 Results

3.3.1 Initial validation

In order to demonstrate the simulation's convergence characteristics, a time course from initial state to convergence is traced in Fig. 3.7, where the monocular excitatory fields M and the stereoptic excitatory fields S are displayed as three-dimensional grids. Inputs were as shown in Fig. 3.5. The retinal position q versus disparity d coordinate system is indicated on the upper left-hand grid in each figure. Activity in the corresponding inhibitory pools U (monocular) and V (binocular) as a function of retinal position q is displayed in the form of line graphs just below each column of two grids. The displays are temporally spaced at intervals of 2 time-constants of the excitatory field and thus represent a total of 10 time-constants of simulated activity. Although this time cannot be directly tied to actual processing time in neural tissue, the benchmark of 10 time constants serves as a basis for comparison with the behavior of the model when tested later under purely monocular stimulation. The initial lack of excitation in the neural fields is depicted in Fig. 3.7a. The gradual buildup of excitation and the corresponding response in the inhibitory pools may be noted in Figs. 3.7b - c. In Fig. 3.7d the growth of inhibition has begun to shrink the widths of the excited areas of the fields. Figs. 3.7e - f show further steps towards reaching a satisfactory depth segmentation. The fields driven monocularly via accommodation cues and binocularly via disparity cues tend towards virtually identical states. Field excitation is maintained only in narrow bands along the disparity axis but shows significant spread in the retinal angle direction as nearby elements reinforce each other. At all excited retinal positions, the potential in the inhibitory pool reaches a level high enough to suppress exitation for all but one disparity.

The final depth segmentation shown in Fig. 3.7f was used to produce the reconstruction shown in Fig. 3.8, which illustrates the result of transforming the equilibrium state of the monocularly driven field back into

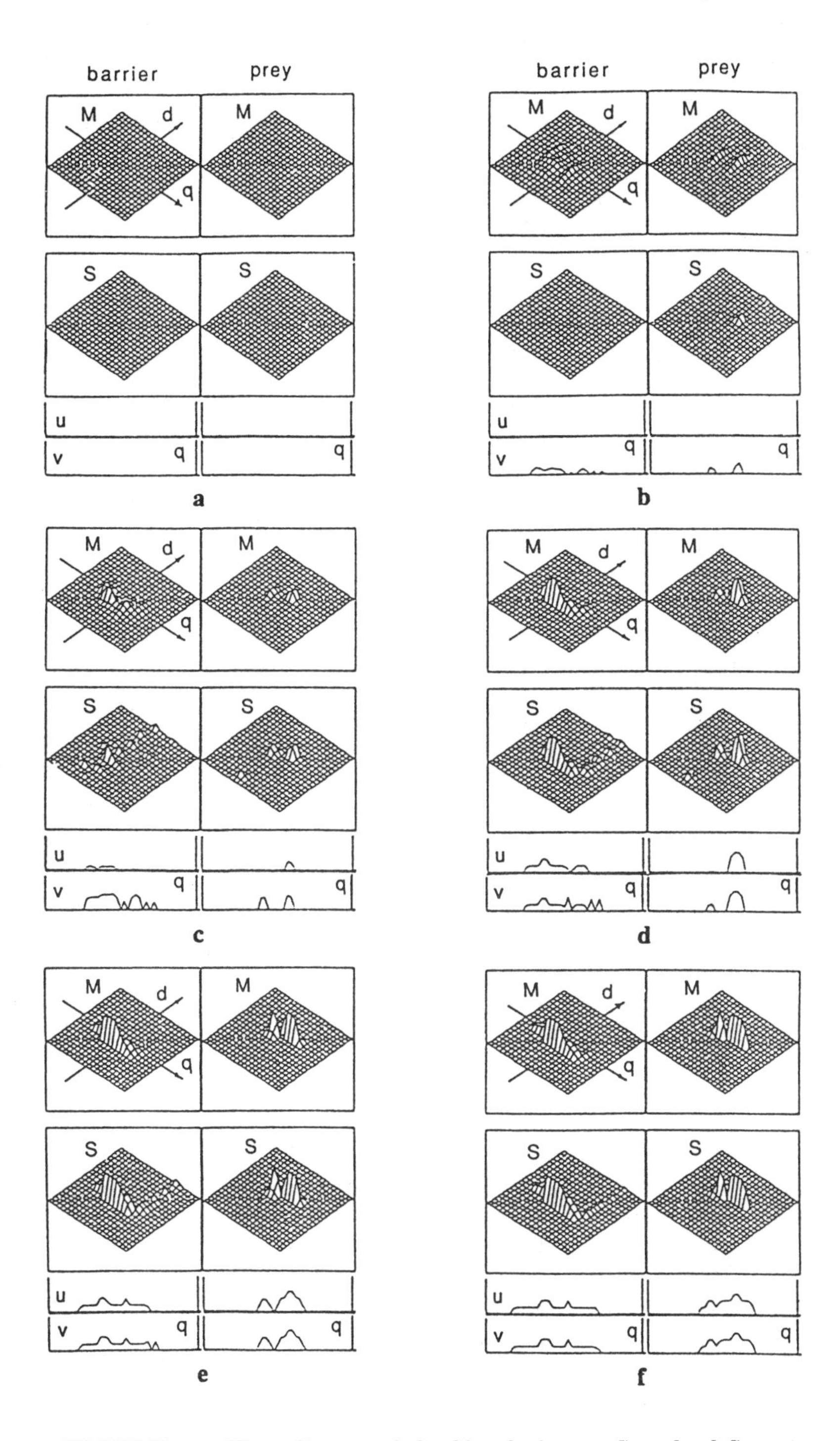

FIGURE 3.7. Time Course of the Simulation — Standard Scene

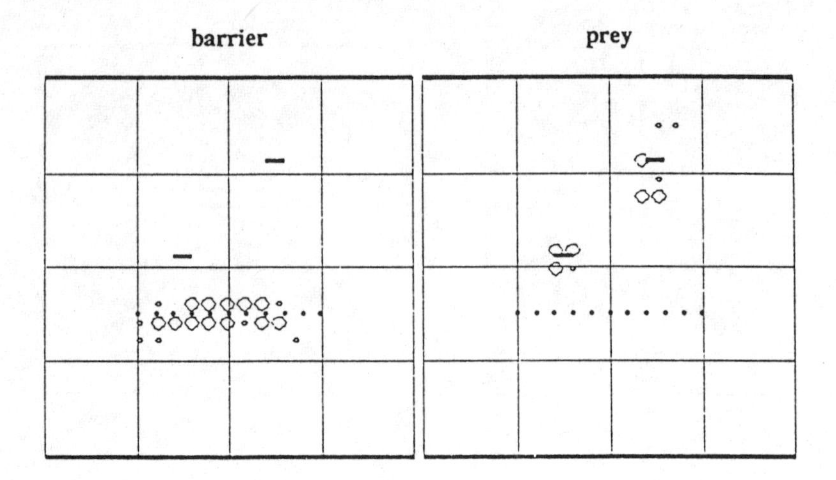

FIGURE 3.8. Scene Reconstruction from Depth Segmentation

Cartesian coordinates. This display is shown overlaid upon the test scene and indicates the extent to which the model was successful in constructing a depth map of its spatial distribution. The decrease of depth acuity with increasing distance may also be noted by comparing the estimates for the two prey objects.

3.3.2 Effect of lateral excitatory spread

Both the performance and the time-to-convergence of the simulation are strongly affected by the strength of the lateral connections in the excitatory fields. A series of runs was made in which the net gain of the spread function w was varied from its nominal setting w_0. Although eqs. (3.1) show two spread functions, w_s and w_m, we kept these functions identical in all of the reported simulations. Here we use the symbol w as a "shorthand" to denote both of these functions. During these runs, performance characteristics and time to convergence were judged visually by watching a graphic display of the simulation. Fig. 3.9 plots observed convergence time as a function of the net strength of the excitatory spread when the standard test scene was provided as input. Four regions where performance was judged to be qualitatively different are indicated at the bottom of this figure. In region I, the excitatory gain was so low that convergence was slow, and no lateral spread of excitation beyond a given point of stimulation was noted. Excited points in the field remained below saturation level, with the level of excitation determined mainly by the strength of the input at a given point. In region II, overall performance was judged to be best. Convergence speed was near optimal and excited points tended to be in saturation. However, in this re-

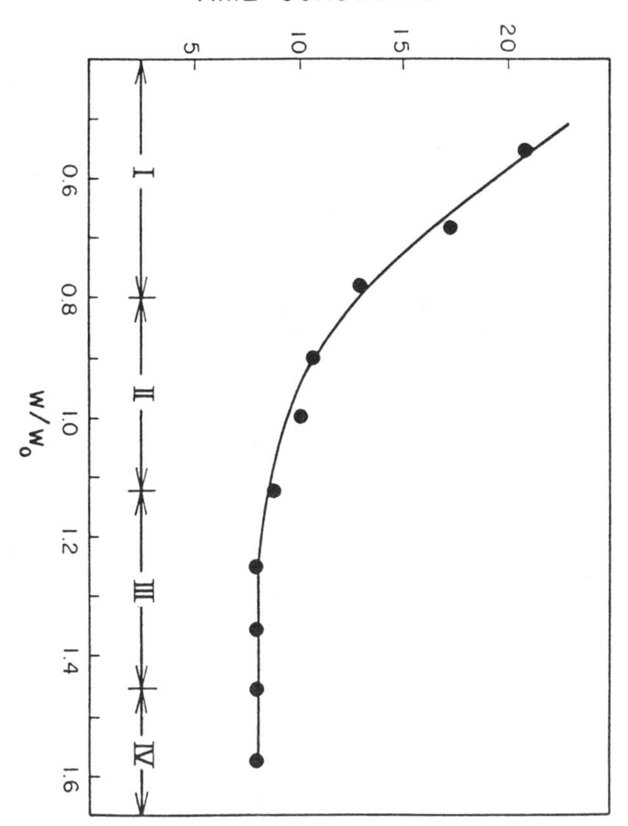

FIGURE 3.9. Convergence Time vs. Strength of Excitatory Spread
The horizontal axis is the net gain w of spread-function, normalized to the nominal gain w_0 used for other experiments. The vertical axis is time-to-convergence, measured in excitatory-field time constants, when the standard test scene is used as input. Regions I to IV indicate gain settings for which performance was judged to be qualitatively different.

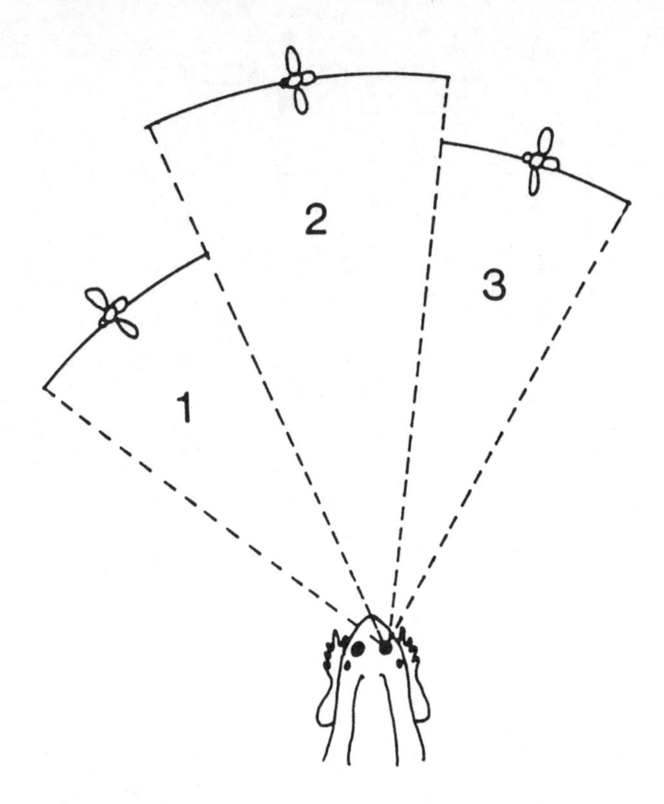

FIGURE 3.10. Visual Field Depth Segmentation Produced by the Model

gion excitation was high enough to spread laterally to points not receiving external stimulation. Convergence was most rapid in regions III and IV, but here the uniqueness constraint (1) was often seen to be violated. In region III, neighboring points at the same retinal position often remained simultaneously excited, thus degrading depth resolution. In region IV, there was a more dramatic breakdown of uniqueness, with multiple disparities at a single retinal position being above threshold.

3.3.3 Depth segmentation

Because of the relationship between model performance and excitatory spread, it is not possible to tune the model for rapid convergence without losing the ability to interpret its output as a pointwise reconstruction of the visual scene. With the model tuned for rapid convergence, the spread of excitation caused by the lateral connections in the cooperative fields results in a segmentation of the scene into depth regions (Fig. 3.10). The breadth of each segment is dependent upon the length of the simulation run and the placement of objects in the scene. Segments grow out laterally on either side of a point of stimulation. Growth of a segment is constrained only by the physical boundaries of the excitatory field or by intersection with a segment growing out from a different point of stimulation. If the model were tested against visually rich stimulation, such as a random-dot stereogram, the intersection of segments would be ubiquitous and segments would be prevented from growing much beyond the range of initial stimulation. The segmentation effect is seen only because of the sparsity constraint (5). Because of the segmentation effect, the spatial coordinates of a visual object cannot be determined by means of the model's output alone. However, spatial location can be determined by the extraction of visual angle from direct retinal stimulation and by sampling the depth map at this angle to determine depth. The segmentation effect has important behavioral implications to be discussed later.

Fig. 3.11 provides an illustration of the depth segmentation effect and demonstrates how the resulting problem of poor angular resolution can be overcome. In order to dramatize the segmentation effect, the simulation was was allowed to run twice as long as the simulation of Figs. 3.7 and 3.8 (20 field time constants). The final state of this run is shown in Fig. 3.11a, and its transformation back into external coordinates is shown in Fig. 3.11b. Finally, Fig. 3.11c shows that, even though the depth segments are greatly exaggerated, objects can still be successfully localized by sampling the depth map at only those retinal positions currently receiving stimulation.

3.3.4 Effects of Lenses and Prisms

Fig. 3.12 shows the depth maps produced by simulation runs employing the same input scene but using simulated interposed lenses (Fig. 3.12a) and prisms (Fig. 3.12b). The simulated lens and prism strengths were chosen to produce a 20% shift in the disparity coordinate of the corresponding monocular (lenses) or binocular (prisms) input plane. The results for prey stimuli show that the shift in the depth segmentation produced by lenses is quite small compared with that produced by prisms. The large shift apparent in the prey depth-estimate using prisms compared with that obtained using lenses is consistent with behavioral data. The different lens and prism results obtained for barriers are explained by the close spacing and spatial periodicity of the fenceposts. Currently, no behavioral data exist for this type of stimulus.

The biasing of the model to favor binocular cues in determining depth of prey is very sensitive to the relative strengths of the binocularly and monocularly derived inputs. Experimentation showed that no internal model parameters could affect this biasing more than the relative weighting of the input gains K_a and K_d. In order to show this result more explicitly, a series of experiments was run with an input scene consisting of a single prey object placed in the center of the scene, 22 cm from the simulated animal along its midline. With lens strength held constant at 20%, the ratio of

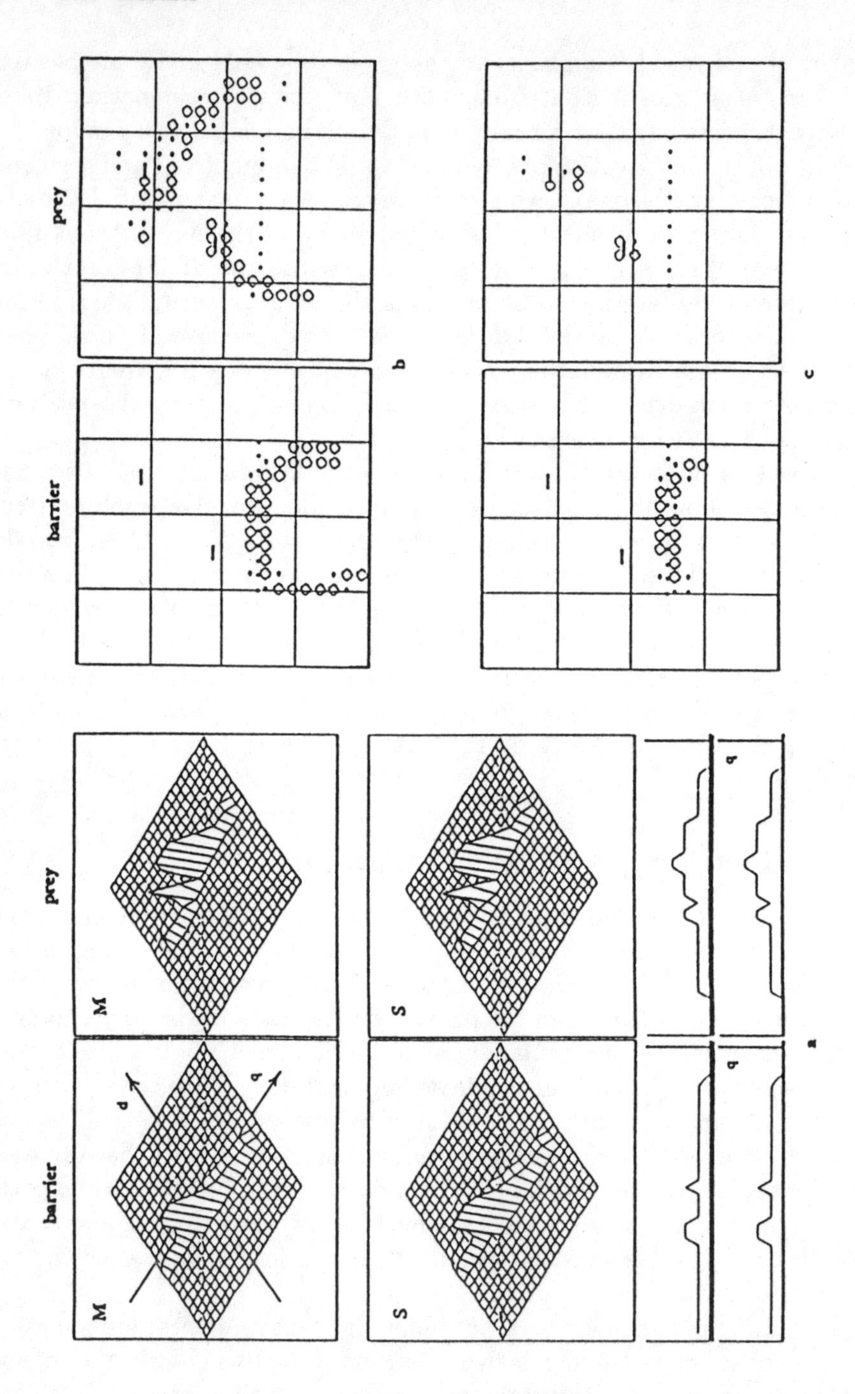

FIGURE 3.11. Object Localization from Depth Segmentation

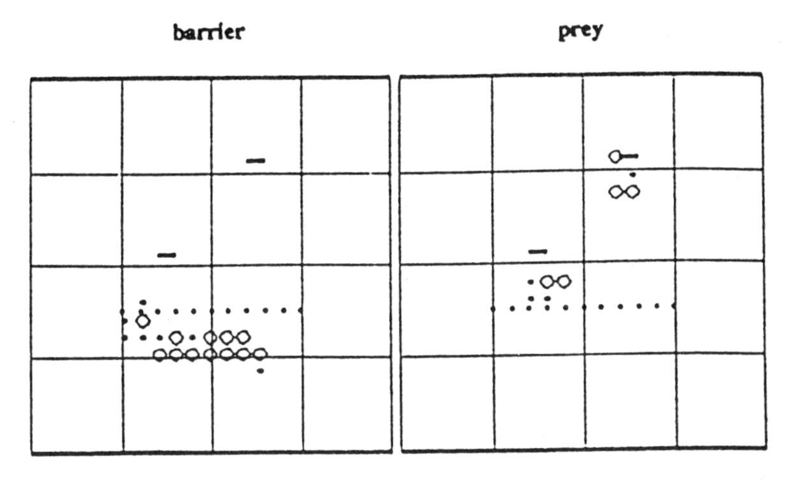

a - lens

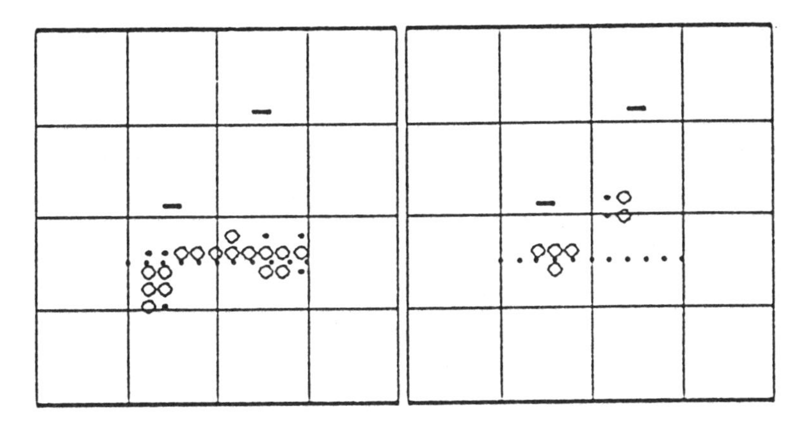

b - prism

FIGURE 3.12. Lens and Prism Effects

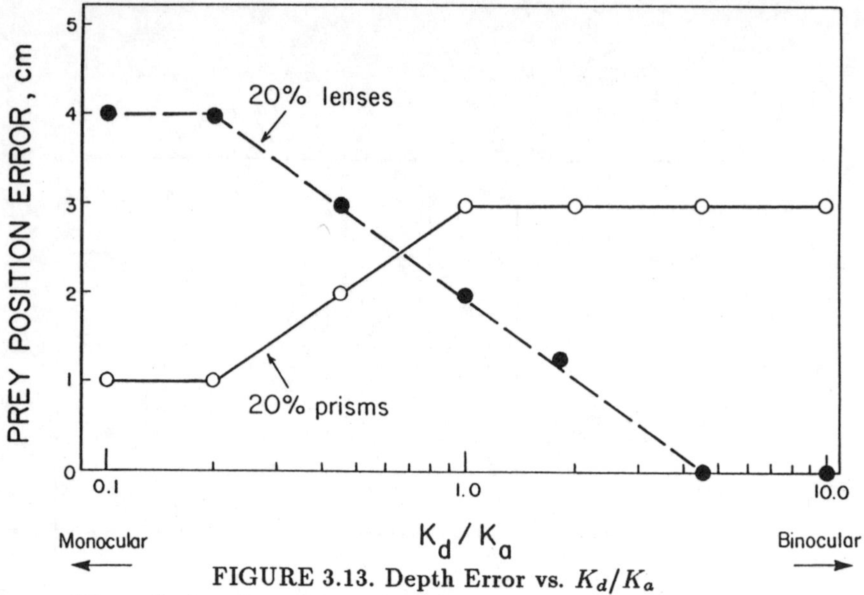

FIGURE 3.13. Depth Error vs. K_d/K_a

Error of the calculated position of a prey object vs. the ratio of the strength of disparity cues K_d to accommodation cues K_a when viewed through 20% lenses (dashed line) or 20% prisms (solid line).

binocular input strength K_d to monocular input strength K_a was varied over two log units (from 0.1 to 10.0) in such a manner as to hold the net input strength $K_d + K_a$ constant. Error in prey depth estimate as a function of this ratio is plotted in Fig. 3.13 together with results using 20% prisms. Setting the ratio K_d/K_a to about 3.0 will reproduce the results obtained by Collett [1977], showing a major binocular effect and only a minor monocular effect. The steep slope of the prism curve compared with the more shallow slope of the lens curve can be accounted for by the difference in acuity of binocular versus monocular cues.

3.3.5 Barrier depth resolution

Depth resolution of prey in the presence of lenses and prisms is consistent with experimental results. There have been no experiments on the resolution of barrier depth, so the results of our simulations may be seen to yield interesting predictions. In the simulations depicted in Fig. 3.12, the lenses produced a large shift in the depth segmentation whereas the shift due to prisms was smaller and in the opposite direction. This effect is further explored in the experiments depicted in Fig. 3.14, which shows the depth analysis for a single centrally placed fence and a range of lens and prism settings. The effect of lenses is to shift the depth estimate closer to the animal as the lens strength increases. However, the effect of increasing prism strength is less consistent. Depth estimates shift forward and back,

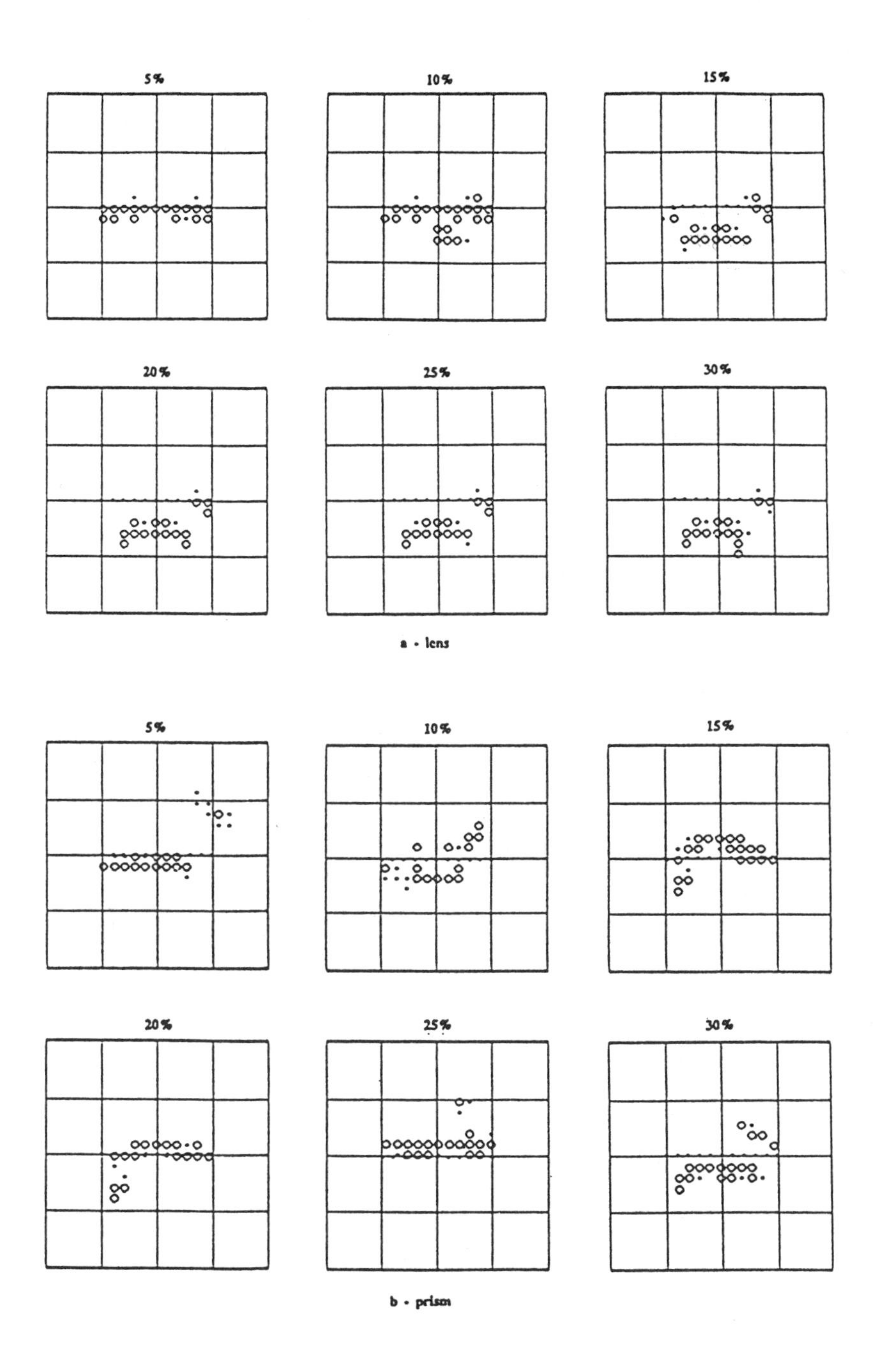

FIGURE 3.14. Effects of Lenses and Prisms on Fence Depth Estimation

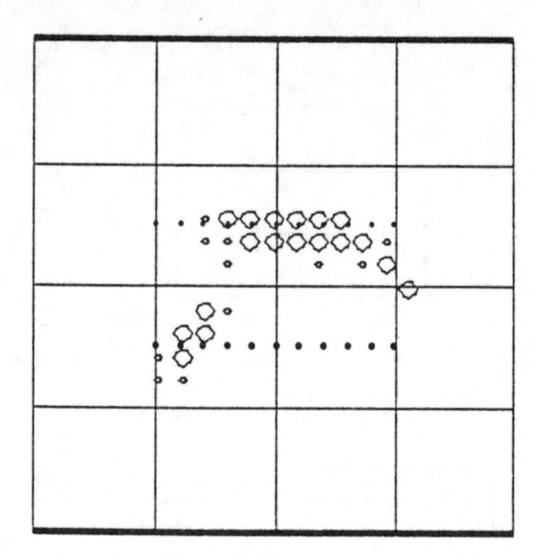

FIGURE 3.15. Depth Estimation With Two Fences

often fragmenting into several depth regions. This effect can be explained by the close disparity spacing of the spurious binocular depth estimates derived from the repetitive fenceposts. Any mismatch between monocular and binocular estimates that exceeds this disparity spacing leads to a reinforcement of one of the spurious estimates rather than to a reinforcement of the correct one. For more visually sparse data, such as prey stimuli, reinforcement of spurious estimates is unlikely.

Another experiment was done to examine the response of the model when confronted with two fences. Fig. 3.15 shows the model's response to a scene containing two fences placed one in front of the other. The simulation resolves portions of each fence but does not completely resolve both. This is because, in conformance with the uniqueness constraint (1), the model was tuned so that excitation could not be sustained at multiple depths for a single retinal position.

3.3.6 Monocular Response

A final experiment was done to test the model's performance when deprived of binocular depth cues, thus replicating a monocular animal. Fig. 3.16 shows the results of this experiment. The figure shows that with the binocular input strength K_d set to zero the model can still produce a correct depth segmentation. Although accuracy was not noticeably impaired in this simulation run (c.f. Fig. 3.8), the length of time required to reach a satisfactory segmentation was doubled, from 10 to 20 field time-constants. The slowdown in convergence time is due to the net reduction of input gain incurred by removing one of the model's input sources.

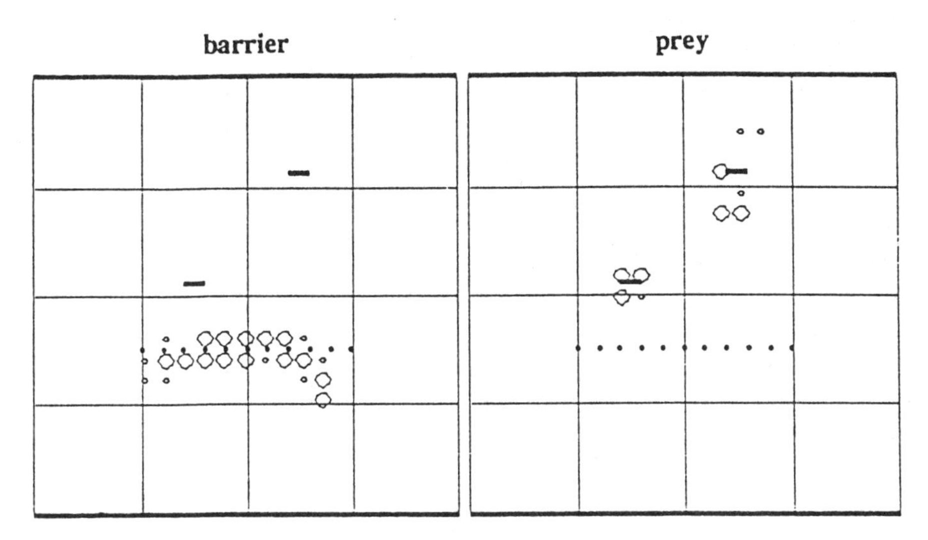

FIGURE 3.16. Monocular Response

3.4 Discussion

One of our initial goals was to design a stereoscopic depth-perception model that conforms to the methodology used in previous cooperative models, but that employs accommodation cues, instead of cues from vergence, to resolve binocular ambiguity. A second goal was to define a model that will result in depth maps dominated by binocular depth cues, but that can still function when only monocular cues are available. Simulation runs have shown that the model is successful in meeting both of these goals. Since disparity matching is not confined to a small area around a point of stimulation, vergence cues do not play a role in the depth mapping process. Instead, the range of disparities to be considered is effectively constrained by accommodation cues. Disparity cues that match strong accommodation cues are much more likely to be selected than are those that do not. Also, the simulation can be tuned to provide an adequate depth mapping both with and without binocular information, and under this same tuning its depth mapping process is dominated by binocular cues when they are present.

The model differs from previous models in its use of accommodation to obtain depth cues. Earlier models were tested with only static stereoscopic images as input. Since our model uses the process of accommodation to obtain secondary depth cues, it must be tested against a model of the input scene that includes the depth dimension. In this sense, it is more suitable for representing the depth resolution process of animals. For this same reason, its method of using accommodation cues to disambiguate binocular cues might be useful in a robotic depth perception system. Robots interact, not with static images, but with true three-dimensional scenes, and thus could benefit from the use of depth cues obtained from accommodation.

Inspiration for the model and tuning data for the simulations were taken from the literature on the visual system of frogs and toads. In particular, tuning of the relative weights of accommodation and disparity cues was guided by consideration of data on prey-catching in toads [Collett, 1977]. We confirmed that the simulation can be tuned so that its discrimination of the spatial locations of prey-like stimuli is only slightly affected by the presence of moderate systematic error in the accommodative mechanism (lens effect), while remaining very sensitive to systematic binocular error (prism effect). Since the ratio of accommodation to disparity cues used to produce these results is only one of many possible choices, the model can be thought of as representing a spectrum of depth perception mechanisms that range from one totally dependent on accommodation cues to one that uses only disparity cues.

The model also leads to new predictions concerning the errors that frogs and toads might make in determining the depth of fence-like barriers. When both the monocular and the binocular mechanisms are perfectly in tune with each other, fence depth is correctly ascertained. However, a small detuning of either mechanism will lead to a breakdown in consistent discrimination capability. Further, there is a qualitative difference between the effect of lenses and the effect of prisms on this type of depth resolution. Lenses have the expected effect in shifting the depth estimate. However, prisms have a less predictable effect and tend to produce an image fragmented into several depth regions.

The use of accommodation to disambiguate, and not merely supplement, binocular estimates resolves a problem posed by Collett and Harkness [1982]. Their calculations show that in small animals the use of a weighted sum of binocular and monocular depth cues, rather than purely binocular cues, can produce only a small improvement in accuracy. For instance, in toads, the use of such a weighting scheme can yield only a 5% improvement. It is not clear, based on this analysis, what advantage frogs and toads gain by the simultaneous use of both types of cue. However, our analysis shows that if accommodation cues are used to help select among competing binocular cues, and not merely to refine accuracy, depth perception can be accomplished with greatly enhanced speed and reliability.

In developing the model, we assumed that frogs and toads are able to simultaneously maintain separate depth segmentations of prey and barrier features within the visual field. We also assumed that direct optic-nerve input provides the angular position of an object, while a depth segmentation map provides its depth. Arbib and House [1983] further exploit these assumptions in the development of models of the orientation and barrier negotiation behavior of frogs and toads.

4

Localization of Prey

ABSTRACT In this chapter we present an action-oriented model of the spatial localization of prey by frogs and toads. Instead of building a global depth map, we propose that the goal of catching prey can lead a frog or toad to select a particular region of its visual world for special scrutiny. We suggest that the first step of the prey-catching sequence is not an overt movement, but a covert movement to adjust the accommodative state of the lenses and thus lock the visual apparatus on to a stimulus. We demonstrate how prey localization can be acheived rapidly and accurately by coupling prey-selection and lens-accommodation processes within a feedback loop. Information derived from prey selection supplies a setpoint for accommodation. In turn, adjustment of the lens modifies the visual input and can alter the prey selection process. The natural feedback of this goal-seeking system automatically corrects for the problem of ambiguity in binocular matching. We tie the model to the known anatomy, physiology, and behavior of frogs and toads, identifying brain regions that could provide the neural substrates necessary to support the model's various functional stages. We also present experiments, with a computer simulation, that compare the model's functioning with animal behavior.

4.0.1 BACKGROUND

To date there is little evidence that frogs and toads have the anatomical and neurophysiological features necessary to support a global depth-mapping process based solely upon binocular disparity. In the previous chapter we investigated a mechanism of cooperativity between binocular and monocular cues to determine depth. The model developed in that study was successful in integrating these cues into a coherent scheme, but it utilized disparity detectors, and its output was a global depth map. The model presented in this chapter maintains the notion of the coupling of binocular and monocular mechanisms. However, it does not depend upon disparity detectors and, rather than computing a global depth map, it localizes only a single point in space.

The finding that toads lacking their major source of tectal binocular information [Grobstein and Comer, 1983] still use binocular depth perception challenged the notion that a depth map could be computed in tectum. In place of local disparity matching, Collett and Udin suggested a process similar to that observed in mantids [Rossel, 1980]. In their scheme, each tectal lobe selects a particular visual object and passes its retinal position to a

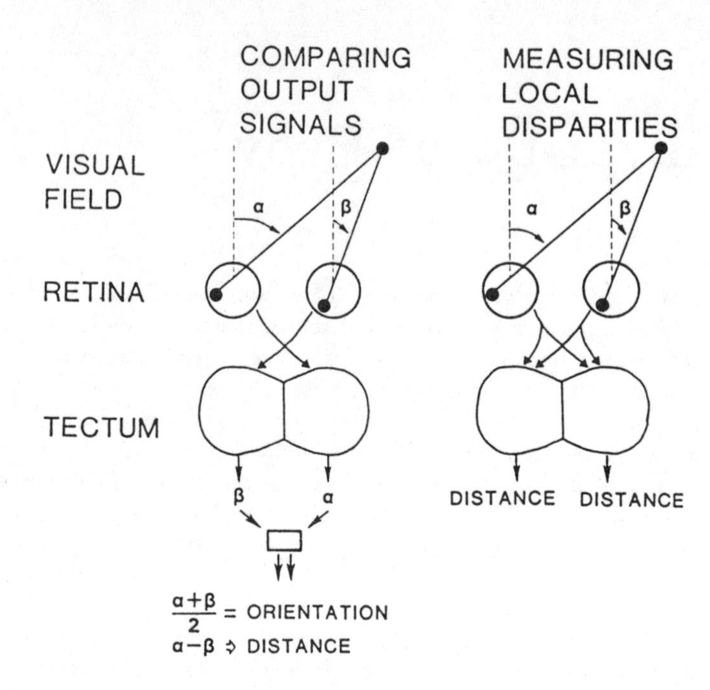

FIGURE 4.1. Depth from Selection vs. Depth from Local Disparity

common brain region. This region then uses these two retinal angles to determine the spatial location of the object. Fig. 4.1 depicts this scheme and contrasts it with one based upon local disparity matching, where both tectal lobes build a depth map by measuring local disparities between points on the two images. Tectal output would then include an encoding of this depth.

Although toads with lesions to the isthmic nuclei are as accurate as controls in striking at a single prey object, they are subject to errors in binocular correspondence not seen in the controls. This result indicates that the role of the isthmic nuclei in this process is to provide cross-tectal relays that assist the two tectal hemispheres in agreeing upon the same visual object. If the isthmic connections are lost, then each tectal hemisphere will independently arrive at its own choice of stimulus and binocular mismatch will be probable.

An apparent weakness of Collett and Udin's model is that, when presented with two prey stimuli, it chooses the prey nearer to the horopter. Their computer simulations showed that if the horopter surface, in relation to which the ipsilateral and contralateral visual fields are in register in tectum, is at 50 cm (for a 3 cm eye separation) then the chance of binocular mismatch is minimized. However, since the snap zone of toads is well within 50 cm [Ingle, 1970], placing the horopter at this distance results in the selection of the more distant of two prey objects. Since actual animals prefer

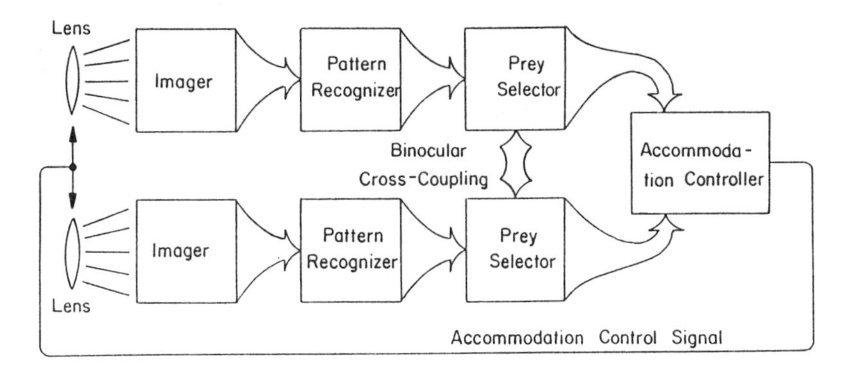

FIGURE 4.2. Functional Diagram of the Prey Localization Model

the nearer of two equally attractive prey objects, this result is inconsistent with behavior. Collett and Udin addressed this problem by suggesting that lens accommodation could play a role in bringing the nearer prey into sharper focus, thus making it more salient.

In the next section we modify and extend the ideas of Collett and Udin by proposing a model in which binocular prey-selection and lens-accommodation processes are tightly coupled in a feedback loop. The coupled system is shown to converge rapidly upon a single prey object and, under a reasonable set of assumptions, will select the nearer of two prey objects.

4.0.2 Model Overview

The functional components of the model consist of a binocular imaging system, pattern recognizers, prey selectors, and an accommodation controller. Fig. 4.2 depicts these components and shows how they are interconnected. The imagers produce an output pattern whose intensity at any position is dependent upon the crispness-of-focus of the image at that position. The pattern recognizers take their input from the imagers and, in turn, provide an output modulated by the degree to which each region of the image matches recognition requirements. It is assumed that the pattern recognition process is optimized when the image is in sharp focus. The outputs of the pattern recognizers project to the prey selectors. The selectors are responsible for identifying the region of the original image that corresponds to the strongest signal from the pattern recognizers. The two prey selectors are also binocularly cross-coupled to assist them in agreeing upon the same visual object. The outputs of the prey selectors are the image coordinates of the selected object. These coordinates project to the accommodation controller where they are used to compute depth from binocular disparity. This depth, in turn, is used to adjust the accommodative state of the lens.

Adjusting lens focus completes a feedback loop that, along with the binocular cross-coupling of the prey selectors, assists in the resolution of binocular ambiguity. If both prey selectors have chosen the same object in real space, then the resulting depth estimate will be correct and the effect will be to bring this object into sharper focus. The improvement of focus will enhance the pattern recognition process and improve both the confidence and angular resolution of the recognizer's output. However, if the prey selectors have chosen different objects, the depth estimate will be incorrect and the lens will be driven out of focus. This will affect activity in the pattern recognizers and their new outputs will alter the selection process.

If only a single object is present in the visual field, there will be no source of binocular ambiguity and the system will rapidly localize this object. If multiple objects are present, but at different depths, the action will be to bring one of these objects into clear focus, and again a correct depth estimation will be assured. The more difficult problem of multiple objects at the same depth will be resolved by the cross-coupling of the prey selectors. Thus, the only situations that will be ambiguous will be those in which the visual angle between two objects, at the same depth, is smaller than the resolution of the binocular cross-connections.

4.1 Methods

4.1.1 Computer simulation

A computer simulation of the model was developed so that experiments could be done to evaluate its performance. This simulation is implemented in Pascal on a VAX – 11/780. Model results are displayed using a 512×512 pixel Grinnell color display. A menu-driven interactive graphics system, using this same display, allows the experimenter to adjust model parameters and test-scene configurations. Hard-copy images mimicking the interactive display are produced on a Symbolics laser-graphics printer by a batch version of the modeling system.

Several simplifications and assumptions were made when constructing the simulation and designing experiments. The test scenes used in all of our experiments consisted of either one or two elongated, textureless stimuli in the frontal visual field. We did not attempt to represent a background. The simulation treats only two-dimensional scenes projected onto one-dimensional image planes. We assumed that the lenses are coupled so that both lenses are always accommodated to the same depth.

4.1.2 Mathematical description for the simulation

The functional elements of the model are represented as layered networks of cellular components. This constraint was adopted so that computational

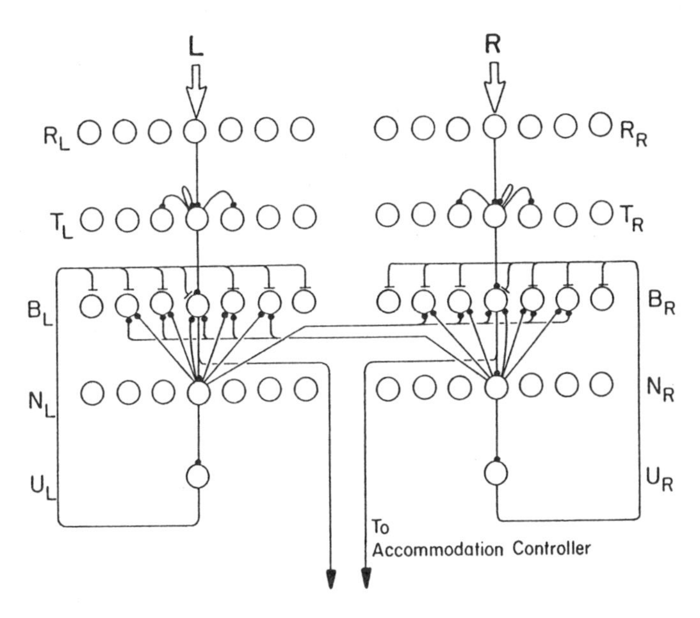

FIGURE 4.3. Implementation Schematic of Prey-Localization Model
Rows of circles represent cell layers. Intercellular connections that end in filled circles are excitatory and those ending in T's are inhibitory. The imagers R_L and R_R supply visual input to the pattern recognizers T_L and T_R The prey selection function is performed by the interaction of binocular layers B_L and B_R relay layers N_L and N_R and inhibitory layers U_L and U_R.

algorithms would be suggestive of possible underlying neural mechanisms. Fig. 4.3 shows these layers and their interconnections. For simplicity, only output connections from the central cell of each layer are shown. Similar connections are made for all other cells in the corresponding layers. Although this diagram shows only seven cells in each layer, the actual computer simulation used 41 cells per layer. The mathematical description of the model portrays these layers of cellular units by continuous analytical equations. This convention was adopted for consistency with the description of the *cue interaction model* of Chapter 3. Refer to Appendix A for a discussion of the development of the mathematical description, and for details of the computer simulation based on this description.

The cell layer representing the imagers simply relays image strength to the pattern recognition layer. The image strength is a function of both the intensity of the incident image and the state of lens accommodation. The effect of lens accommodation on the imagers is represented by varying the intensity of the imager's output signal according to the closeness of the projected point in space to accommodation depth. A Gaussian modulating function is used to represent the effect of accommodation and is described in detail in Appendix B. The acuity of accommodation cues decreases with depth. Thus, as with any focusing system, nearby objects are more accurately located in depth than are far objects.

The pattern recognition mechanism is not modeled in any detail since we were concerned only with its output. The need for pattern recognition was circumvented by limiting the visual input to a few selected points. The cellular elements of the pattern-recognition layer are modeled as making simple lateral-excitatory interconnections. Therefore, neighboring stimulated points tend to reinforce each other so that the intensity of the recognizer's output in response to an object is a function of both the strength of the signal from the imagers and the retinal angle subsumed by the object. The level of excitation in the pattern recognition layers may be represented in analytical form by the integro-differential equations

$$\begin{aligned} \mathcal{T}_t\dot{T}_L(q,t) &= -T_L(q,t) + \int w_t(q-\zeta) f_t[T_L(\zeta,t)]d\zeta + K_{at}A_L[q, D_a(t)], \\ \mathcal{T}_t\dot{T}_R(q,t) &= -T_R(q,t) + \int w_t(q-\zeta) f_t[T_R(\zeta,t)]d\zeta + K_{at}A_R[q, D_a(t)]. \end{aligned} \tag{4.1}$$

Profiles of the internal potential in the pattern recognition layers are represented by continuous functions T_L and T_R, with spatial dimension q indicating retinal angle and t representing time. Internal potential is converted to an external potential (or firing rate) by the saturation-threshold function f_t. The function w_t is a symmetric spread function that describes the extent and strength of the lateral excitatory connections between points in the pattern recognition layers. D_a is the current accommodative state of the lenses, represented on a disparity-based depth scale similar to the scale used in the model of Chapter 3 (see Fig. 3.3b). It is used, along with retinal angle q, to index maps A_L and A_R of input from the imagers. The pattern recognizers take their input from the imagers through gain K_{at}. The rates of change of potential in layers T_L and T_R are governed by the time constant $\mathcal{T}_t$. Output from the pattern recognizers projects to a layer of binocular cells.

The binocular cell layer is the layer that actually accomplishes the prey selection. The output of this layer projects both to the accommodation controller and to a relay layer that, in turn, sends connections back to both the ipsilateral and contralateral binocular layers. Cells in the binocular layer sum both these ipsilateral and contralateral signals over broad receptive fields (nominally 16.9°). The input to the binocular layers from the pattern-recognition layer is summed over a narrower receptive field. Thus the binocular cells are excited by a combination of direct stimulation from the pattern recognizers and indirect binocular input. Stimulation from the pattern recognizers assures high angular acuity, and the binocular input biases the selectors so that they tend to select the same visual object. The binocular layer also receives a global inhibitory signal, proportional to the total excitation over the entire binocular layer.

The combination of excitatory receptive fields and global inhibition produces a network that will suppress activity except at that location receiving the maximum stimulation. This network is similar to one proposed by Didday [1970, 1976] (for prey selection in frog) and formalized by Amari and Arbib [1977]. The binocular layers are described analytically by

$$\begin{aligned} \mathcal{T}_b \dot{B}_L(q,t) &= -B_L(q,t) + \int w_b(q-\zeta) f_b[B_L(\zeta,t)]d\zeta \\ &\quad + I_L(q,t) + I_R(q,t) + K_{tb} f_t[T_L(q,t)] - K_{ub} g[U_L(t)], \\ \mathcal{T}_b \dot{B}_R(q,t) &= -B_R(q,t) + \int w_b(q-\zeta) f_b[B_R(\zeta,t)]d\zeta \\ &\quad + I_R(q,t) + I_L(q,t) + K_{tb} f_t[T_R(q,t)] - K_{ub} g[U_R(t)]. \end{aligned} \quad (4.2)$$

Here, B_L and B_R represent internal potential in the binocular layers, f_b is the saturation-threshold function converting internal potential to external potential, w_b is the binocular layer's spread function, and $\mathcal{T}_b$ is its time constant. Inputs I_L and I_R are from the relay layers. The broad spread function associated with these inputs is lumped with the description of the relay layers (see below). The two other inputs are from the pattern recognition layers T_L and T_R through gain K_{tb}, and from the inhibitory layers U_L and U_R through gain K_{ub}.

The relay layer receives its input from the binocular cells and has two distinct outputs. One set of outputs goes to the inhibitory layer, and the other projects to both the ipsilateral and contralateral binocular cell layers. Projections to the inhibitory layer provide that layer with a signal modulated by total activity across the entire binocular layer. Projections to the binocular layers provide the cross-coupling pathways between these two layers. Since the function of the relay layers is simply to pass information to other layers, their transient characteristics can be lumped with those of the other layers. Thus the equations used to describe the relay layers are equilibrium equations rather than differential equations, and are described by

$$\begin{aligned} I_L(q,t) &= \int w_i(q-\zeta) f_b[B_L(\zeta,t)]d\zeta, \\ I_L(q,t) &= \int w_i(q-\zeta) f_b[B_L(\zeta,t)]d\zeta. \end{aligned} \quad (4.3)$$

In these equations I_L and I_R represent the potential transmitted by the relay layers to the binocular and inhibitory layers. The function w_i provides the broad spread of relay potential as received by the binocular layers. This spread function was applied directly to the relay layers and not to their inputs to the binocular layers, in order to give the experimenter direct access to a potential representing the relay's inputs to the binocular layers.

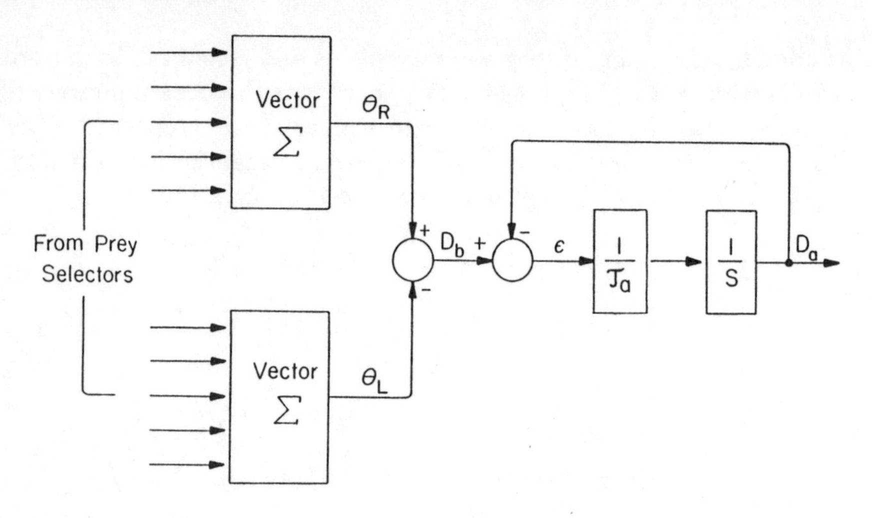

FIGURE 4.4. The Accommodation Controller

The inhibitory layers are single units that simply integrate activity across the entire ipsilateral relay layer. Since the dynamics of the relay layers were not modeled in the simulation, this integration was actually performed over the ipsilateral binocular layer. Thus, the inhibitory layers are described by

$$\begin{aligned}\mathcal{T}_u \dot{U}_L(t) &= -U_L(t) + K_{bu} \int f_b[B_L(\zeta,t)]d\zeta, \\ \mathcal{T}_u \dot{U}_R(t) &= -U_R(t) + K_{bu} \int f_b[B_R(\zeta,t)]d\zeta.\end{aligned} \tag{4.4}$$

Here, U_L and U_R represent the internal potentials in the inhibitory layers, and $\mathcal{T}_u$ is the layers' time constant. The gain K_{bu} is applied to the input from the binocular layers.

The mechanism for adjusting lens accommodative state, based on the output of the prey selectors, is depicted in Fig. 4.4. The first process carried out in this mechanism is to identify a pair of image coordinates by determining the center of gravity, or locus of average excitation, of the activity pattern on each of the two prey selectors. In this process, the output from each point in the binocular cell layer is treated as a vector in a radial coordinate system, with its angular component representing retinal position and its radial component encoding the likelihood that there is a prey object at that position. The vectors from each binocular layer are summed to produce a resultant whose angular component is the retinal position corresponding with the center of gravity on that side of the visual system. The notion, employed here, of motor circuits computing a vector sum of their input in order to control output was originally advanced by Pitts and McCulloch [1947] and was recently confirmed in an analysis of monkey arm movements by Georgopolis et al. [1983], and by McIlwain [1982] in his investigation of cat superior colliculus.

The loci of average excitation θ_L and θ_R in the binocular layers are determined by the formulae

$$\begin{aligned} \theta_L(t) &= \arctan \frac{\int f_b[B_L(\zeta,t)] \sin(\zeta/a) d\zeta}{\int f_b[B_L(\zeta,t)] \cos(\zeta/a) d\zeta}, \\ \theta_R(t) &= \arctan \frac{\int f_b[B_R(\zeta,t)] \sin(\zeta/a) d\zeta}{\int f_b[B_R(\zeta,t)] \cos(\zeta/a) d\zeta}. \end{aligned} \tag{4.5}$$

where the scalar parameter a is simply a conversion factor between visual angle and retinal position. The two retinal angles θ_L and θ_R determined by this vector summation are subtracted to give disparity and thus a measure of depth, according to the formula

$$D_b(t) = a[\theta_R(t) - \theta_L(t)]. \tag{4.6}$$

This depth estimate D_b is the setpoint for the controller. The controller's error signal ϵ is computed by subtracting an estimate of current lens accommodative state D_a from this setpoint. The controller's transfer function is simply a first-order lag with time constant $\mathcal{T}_a$. It is governed by the differential equation

$$\mathcal{T}_a \dot{D}_a(t) = -D_a(t) + D_b(t). \tag{4.7}$$

This scheme for controlling the lens is similar to the efferent-copy scheme that Robinson [1981] utilizes in his model of collicular control of eye movement.

The decision to use an efferent copy, rather than proprioceptive feedback, to provide an estimate of the accommodative state of the lens, was not arbitrary. Toads that have had their lens accommodation muscles treated with a muscle relaxant tend to undershoot prey. When these muscles are treated with a muscle contractor, they tend to overshoot [Jordan et al., 1980]. When the toads' accommodation muscles are relaxed, their lenses are adjusted for far vision. Contraction adjusts the lenses for near vision. Thus if proprioceptive cues were used to estimate lens position, treatment with a relaxant would cause overshooting and treatment with a contractor would cause undershooting. If, on the other hand, an efferent copy were used, then the observed effect would be the expected one.

The accommodation control algorithm is designed to work perfectly when both prey selectors have selected the same visual object. If this has occurred, then the only retinal position excited in each binocular layer will be the one corresponding with the position of the selected prey, and the retinal position from the vector summation will coincide with the position of the prey. Depth can be directly derived from two such retinal positions by a process of triangulation (see Appendix B).

However, before the prey selectors have produced such a refined output, the pattern of excitation over the binocular layers is usually more complex. In this case, the desired depth, computed from the two centers of gravity, will probably not correspond with the actual depth of the true prey object. It will only be an approximation. This approximate depth will generally help to improve focus and, in turn, assist the prey selectors in converging on a single position.

The accommodation controller also contains a refinement that is not depicted in Fig. 4.4. The magnitude of each vector sum is compared with a small threshold. If either magnitude is below this threshold, then the output of the corresponding prey selector is assumed to be too weak to be used to specify a retinal angle. In this case, the desired depth setting of the controller is set to a neutral or rest position.

4.1.3 Graphical displays

The two types of display used to show model output are depicted in Fig. 4.5. Fig. 4.5a is a schematic top-view of a 40 × 40 cm. square arena with grid markings spaced at 10 cm. Below the arena is an icon representing a frog or toad. Filled discs represent the eyes. The filled rectangle in the grid represents a prey object, and the lines projecting into the grid from the centers of the eyes indicate the attention angles calculated by the model. The intersection point of these lines corresponds with the depth being estimated by the binocular system. In this simple test case, the model has correctly located the prey.

Fig. 4.5b depicts the internal state of the model at the same time. The square planes at the top of the figure are maps relating imager output to the possible accommodative states, of the left A_l and right A_r lenses. Their coordinates are visual angle θ, along the horizontal axis, and disparity D (a non-linear representation of depth), along the vertical axis. The circles within these squares indicate, by their size, the strength of the retinal signal that would be produced by accommodating the lens to the depth indicated on the vertical axis. The narrow horizontal rectangles superimposed on these maps indicate the current accommodative state D_a of the lenses. Thus, the area abutting these rectangles represents the current state of the imagers. The three pairs of line graphs below these maps indicate the levels of excitation in the three other layers of the model. The top curves show activity in the pattern recognizers T_l and T_r, the middle curves show the strength of the feedback to the binocular layers from the relay layers N_l and N_r, and the bottom curves show activity in the binocular layers B_l and B_r.

Any binocular imaging system is subject to making errors in binocular correspondence under conditions that vary according to the algorithm being employed. The prey-localization model is no exception to this rule. Since this model uses lens accommodation to help disambiguate binocular

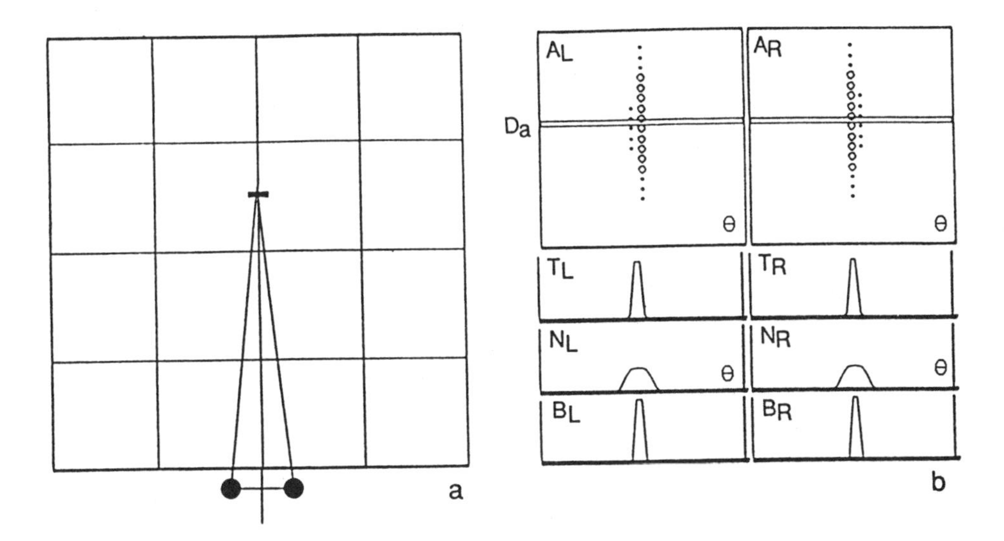

FIGURE 4.5. Displays Available from the Computer Simulation
(a) This display shows both a top-view of the scene being processed and the resulting depth-estimate. The square grid represents a 40 × 40 cm arena. Each grid element is 10 cm on a side. The simulated frog or toad is indicated by the icon below the arena. Filled circles on the cross-bar indicate eye position. The filled rectangle within the arena represents a prey object. The lines emanating from the eyes and converging on the prey indicate the current attention-angles of the two eyes. The point of intersection of these lines determines the model's current estimate of the spatial location of the prey. (b) Squares A_l and A_r represent the full range of possible accommodative states for the left and right eyes. Their vertical coordinates are depth (on a non-linear disparity scale) D, and their horizontal coordinates are retinal angle θ. The long narrow rectangles superimposed on these squares indicate the current accommodative state D_a, of the lenses. The circles within the squares indicate both retinal angle and strength (size-encoded) of a visual stimulus for each accommodative state. The three line-graphs below these squares show activity in the pattern recognizers, T_l and T_r, the relay layers, N_l and N_r, and the binocular prey selector layers, B_l and B_r. Their vertical axes indicate level of excitation and their horizontal axes retinal angle.

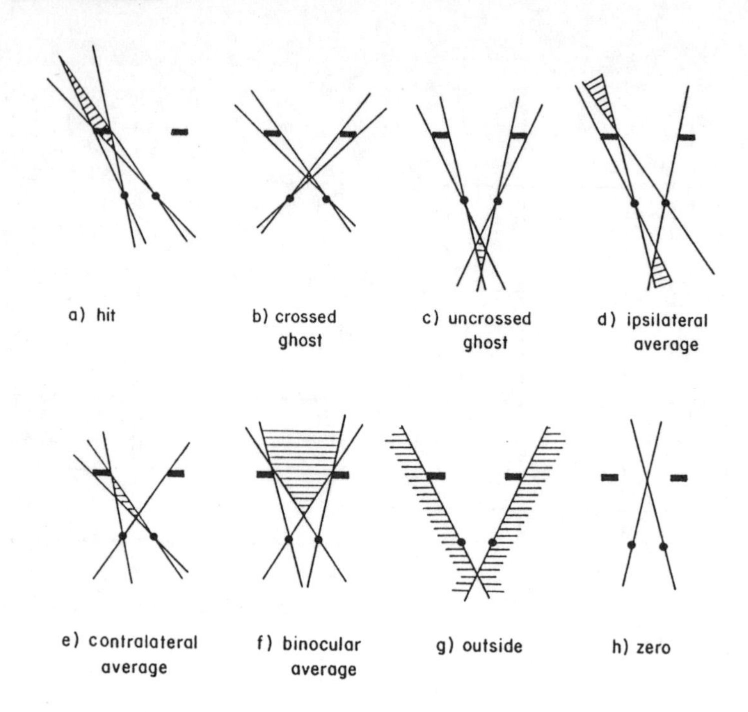

FIGURE 4.6. Categories of Depth Estimates
Dots represent pupil position and filled rectangles represent prey.

matching, it is particularly subject to errors when presented with a visual configuration that is symmetric about the body axis of the simulated animal. Several of our experiments were directed at determining parameter settings that would minimize binocular mismatch under these conditions.

4.1.4 Depth estimate categories

A depth-resolving system based on binocularity will make a correct depth estimate when like points on both retinal images are paired but will make errors when points are incorrectly paired. In order to provide a consistent means for measuring model performance when it was tested with symmetrically balanced visual input from two prey objects, we defined a set of categories of depth estimates. Fig. 4.6 depicts these categories. Since each side of the visual system selects a particular visual angle, the depth estimate is the point of intersection of rays traced from the selected retinal point through the pupil. The hatched area represents the locus of depth-estimates that would result from binocularly pairing angular orientations from within the indicated ranges. Category a) of Fig. 4.6 consists of the locus of points in space corresponding to a *hit* or a correct estimate. This will occur when both rays pass through the same target. Category b) errors represent the binocular or *crossed-ghost* error usually described in the literature. This error occurs when each eye selects the target in its contralateral field. Category c) errors, which we call the ***uncrossed-ghost*** error, represent another true mismatch error where each eye selects the object in its ipsilateral field. As long as the objects in the field are separated by more than the interpupillary distance, this kind of error will result in depth estimates that lie behind the eyes. Categories d) through f) correspond with cases where one or both eyes select a point between the two visual stimuli. Such off-target selection could easily occur in the brain if areas of excitation spreading simultaneously from two stimulated loci interact to create a false peak of excitation between the original stimulus sites. We have labeled these categories (d) the ***ipsilateral-average*** error, where one eye selects the ipsilateral object but the other is directed toward a point between the two objects; (e) the ***contralateral-average*** error, where one eye selects the contralateral object but the other is directed toward a point between the two objects; and (f) the ***binocular-average*** error, where both eyes select a point between the two objects. Category (g) or ***outside*** errors are made when at least one eye selects a point that does not lie either on or between the two stimuli. Finally, for completeness, we include the *zero* error that occurs when neither eye selects any point, leaving the animal's attention directed at a neutral or "relaxed" point.

4.1.5 VISUAL INPUT

Experiments to evaluate model performance with symmetrical input consisted of a series of 21 runs. The visual input supplied for these runs was two identical objects placed symmetrically about the midline and at equal distances from the eyes. To generate the 21 different cases, the objects were placed at 10, 20, and 30 cm from each other and at 7, 12, 17, 22, 27, 32, and 37 cm from the interpupillary line. Results from these runs were tabulated in the form of histograms, with one bin for each of the 8 categories of depth estimate. When runs were made with the intent of performing a sensitivity analysis, the model was run on all 21 cases for each of 6 values of the parameter being studied. Thus histograms contain 6 bins for each category for a total of 48 bins.

4.2 Results

4.2.1 EXPERIMENTS WITH SINGLE-PREY STIMULI

Experiments with a single prey-stimulus were done to verify that the model would function as expected, to tune the model's parameters, and to test the effects of simulated lenses and prisms in the visual field. Fig. 4.7 shows typical results from these experiments.

When lenses and prisms were not used, the model converged rapidly and

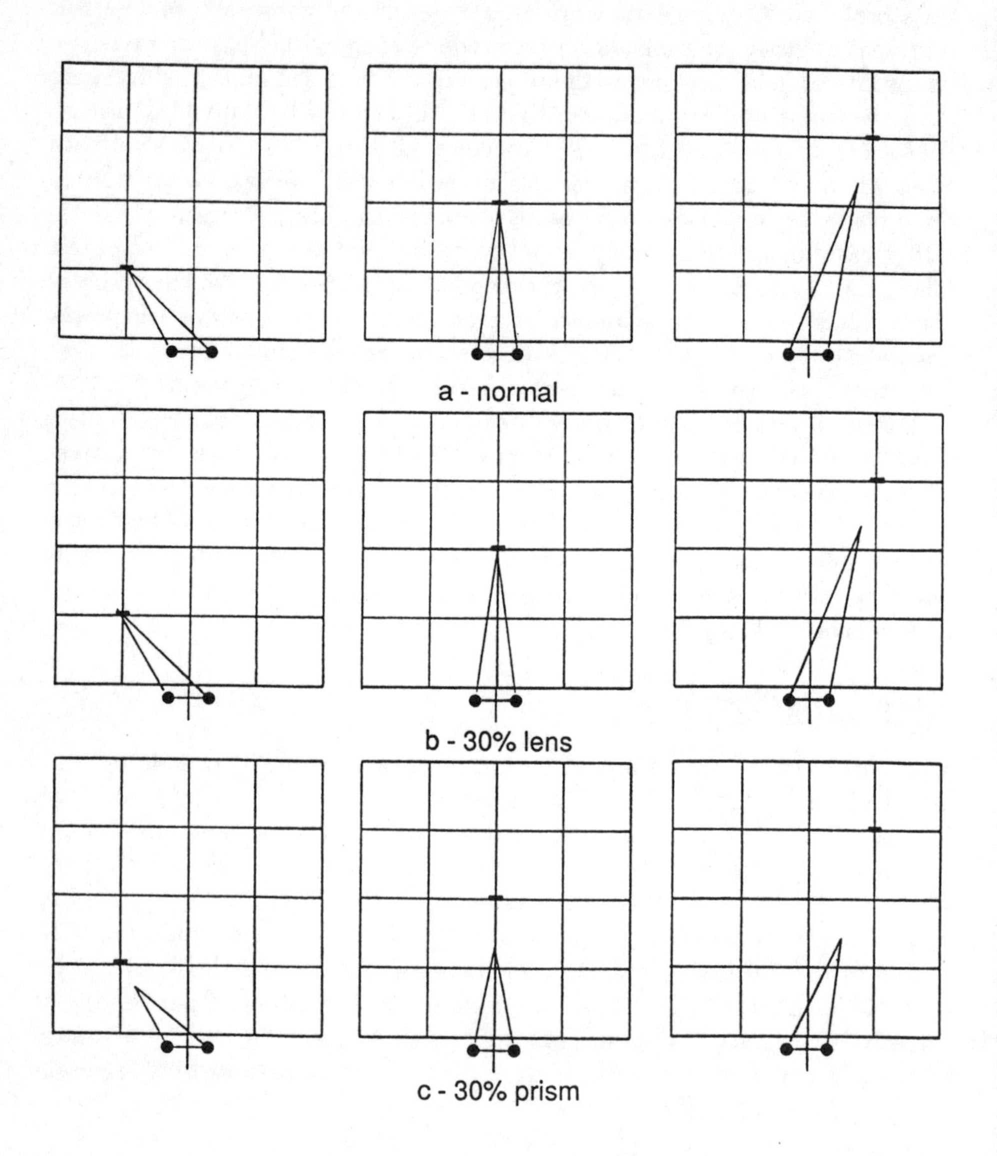

FIGURE 4.7. Sample Results for a Single Prey Object
a) The model correctly localizes a single prey object at various positions. b) Simulated lenses producing a 30% shift in accommodative depth estimate have no effect on single-prey localization. c) Simulated prisms, producing a 30% shift in binocular parallax, cause undershooting errors.

accurately upon a single stimulus wherever it was placed in the binocular field. Fig. 4.7a depicts three such experiments. The reduction in accuracy for the most distant prey is within the bounds determined by the grain-size of the simulation.

When only a single prey object was present, the presence of simulated lenses did not affect accuracy but did increase convergence time. With simulated lenses set to produce a 30% shift in the disparity coordinate of the peak of the lens accommodation curve, the average convergence time for eight different single-prey presentations was 1.8 times that obtained without lenses. Fig. 4.7b shows typical results for the trials with lenses. Accuracy was not affected by the lenses since accommodation was not used as a depth estimator but only to disambiguate binocular matching. With only a single stimulus, there was no ambiguity. The convergence time increased since the defocusing of the lenses reduced the level of excitation in the pattern recognizers and thus acted to reduce the total gain of the system.

Because the depth estimate computed by the model is based solely on binocular parallax, placing base-out prisms in front of the eyes caused undershooting errors. Fig. 4.7c illustrates this effect. Here, simulated base-out prisms, set to produce a 30% shift in the disparity coordinate, caused the expected underestimation of prey distance. Again, the discrepancy between the binocular depth estimate and the lens accommodative state reduced the system gain and resulted in a convergence time 1.8 times greater than that obtained without prisms.

4.2.2 Experiments with Multiple-Prey Stimuli

Visual configurations consisting of more than one stimulus present a correspondence problem that any binocular depth resolution system must be equipped to solve. In our model, lens accommodation provides the additional information necessary to assure correct correspondence. Lens accommodation will be most effective in disambiguating binocular matching when the visual stimuli are at different depths. To confirm that our model is successful in correctly interpreting this type of configuration, we examined its response to several two-prey configurations where the prey objects were placed at different distances from the simulated frog. Fig. 4.8 depicts typical results from a series of such experiments. The model converged upon a correct depth estimation in all but three out of 120 trials. Of the three incorrect estimates, two were *uncrossed-ghost* errors and the third was a *zero* error. In all correct cases, the selected prey was the one closer to the animal. The simulation was biased toward the nearer of the two prey objects by the lateral facilitation in the pattern recognizers. This caused their output to be modulated by the retinal angle subsumed by the visual stimulus. Thus not only are the mechanisms envisioned in the model successful in overcoming ambiguity in binocular matching but they faithfully mimic

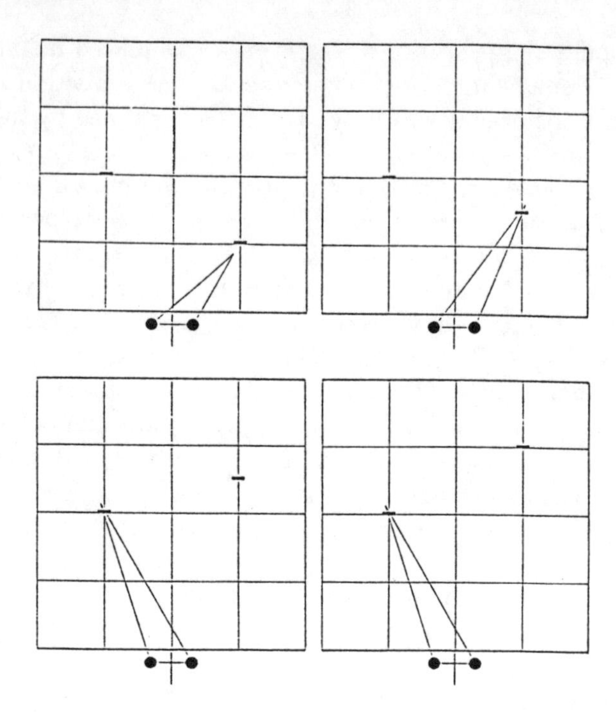

FIGURE 4.8. Results with Two Prey Objects at Different Depths

the demonstrated preference of frogs and toads for the nearer of two prey objects.

A much more difficult problem for our model arises when two prey objects are placed at identical depths. In this case, the use of accommodation cues will not assist in disambiguating the binocular matching problem. However, the crossed pathways between the prey selectors are effective, in most cases, in resolving the ambiguity. Fig. 4.9 shows the range of results obtained during experiments with two prey objects presented symmetrically about the animal's midline. When the model was tuned for optimal performance it produced correct depth estimates for nearly 80% of the test cases. Figs. 4.9a and 4.9b show two representative cases in which a correct depth estimate was made.

Under certain circumstances, binocular cross-coupling is not enough to assure a correct match. The most difficult situation is when the two prey objects are close enough together so that the visual angle separating them is less than the receptive-field size of the binocular crossed connections in the prey selectors. In this case, two distinct kinds of errors can be made. First, the model may make a *ghost* error. If each eye selects the object in its contralateral visual field, then a *ghost* prey will be selected such as that shown in Fig. 4.9c. This *ghost* will be between the actual prey objects and much closer to the animal than either of them. If each eye selects the ipsilateral object then the *ghost* will appear to be behind the eyes, as

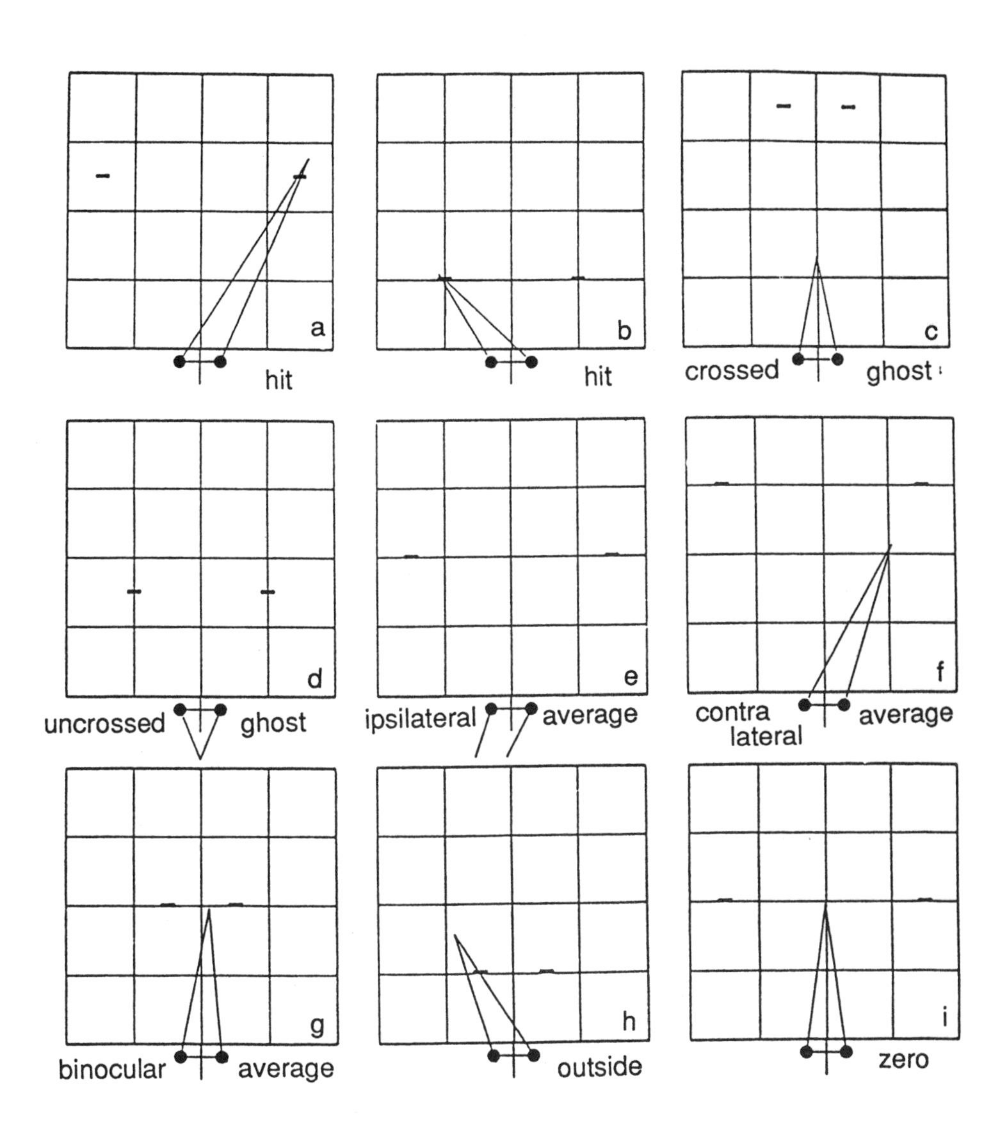

FIGURE 4.9. Results with Two Prey Objects at the Same Depth

shown in Fig. 4.9d. Second, the model may make one of the three *average* errors. In these cases, the model may choose an *average* prey object located at a position between the two true prey objects. The *ipsilateral-average* error is shown in Fig. 4.9e, the *contralateral-average* error in Fig. 4.9f, and the *binocular-average* error in Fig. 4.9g. Besides these *ghost* and *average* errors it is possible to detune the model so that it will locate a prey object peripheral to either of the true prey objects as in Fig. 4.9h, or to ignore both prey objects as in Fig. 4.9i.

The parameters that appeared to have the greatest effect on the response of the model to symmetric presentations were (1) the receptive field width in the binocular layer, (2) the net input gain in the binocular layer, and (3) the time constant of the accommodation controller. A series of three experiments was run to evaluate the effect of varying each of these parameters. For each experiment, the standard test-suite of 21 symmetrical configurations was tried against six different values of the parameter being evaluated. Results were summarized in the form of histograms.

The results of the experiment designed to examine the effect of binocular receptive field width are presented in Fig. 4.10. The six runs in the experiment were done with receptive field set at widths varying from 2.8° to 28.1°. The net gain of the binocular layer's excitatory spread function was held constant by reducing its height in proportion to its increase in width. With the spread set at its lowest value (2.8°) binocular parallax is large enough for most configurations that the binocular fields do not overlap and thus the number of correct matches is only 25%. For all other configurations, the binocular system is essentially disfunctional and the system either does not respond (zero error) or responds with one of the *average* errors. With the optimal setting between 11.2° and 16.9°, responses are nearly 80% correct. The only error occuring in more than one case is the *uncrossed-ghost* error. Increasing the binocular spread to 28.1° results in an increase in both *crossed-ghost* and *ipsilateral-average* errors.

Adjusting the net input gain in the binocular layers has a dramatic effect on the number of correct localizations vs. the number of *average* errors. Fig. 4.11 shows model performance as the binocular gain is varied from 2 to 167% of its nominal setting (run 4) with binocular spread held constant at 16.9°. With the binocular gain very low, there is a preponderance of *binocular-average* errors. This is because the input to the prey selectors is too weak to allow the selection process to isolate a single prey object. The vector summation performed in the accommodation system then simply chooses an average point of stimulation. As binocular gain is increased to near-nominal, the prey selectors begin to operate properly and the number of *average* errors drops off dramatically. Finally, with the binocular gain increased above nominal, the number of *average* errors increases again. The reason for this increase is that excitation is so high in the binocular layers that there is significant spread to lateral positions. Thus the prey selectors cannot successfully select a single visual angle but, instead, select

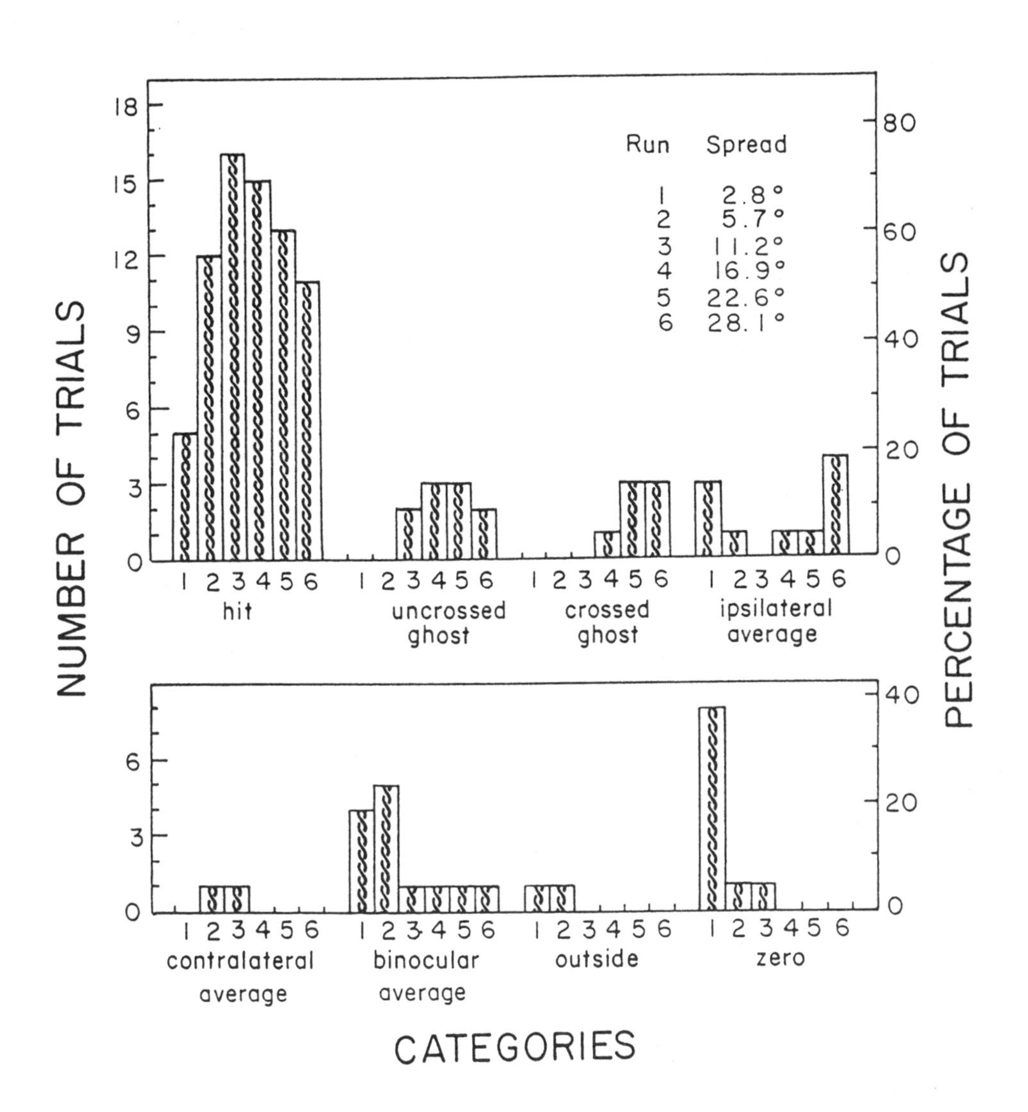

FIGURE 4.10. Model Performance vs. Binocular Spread

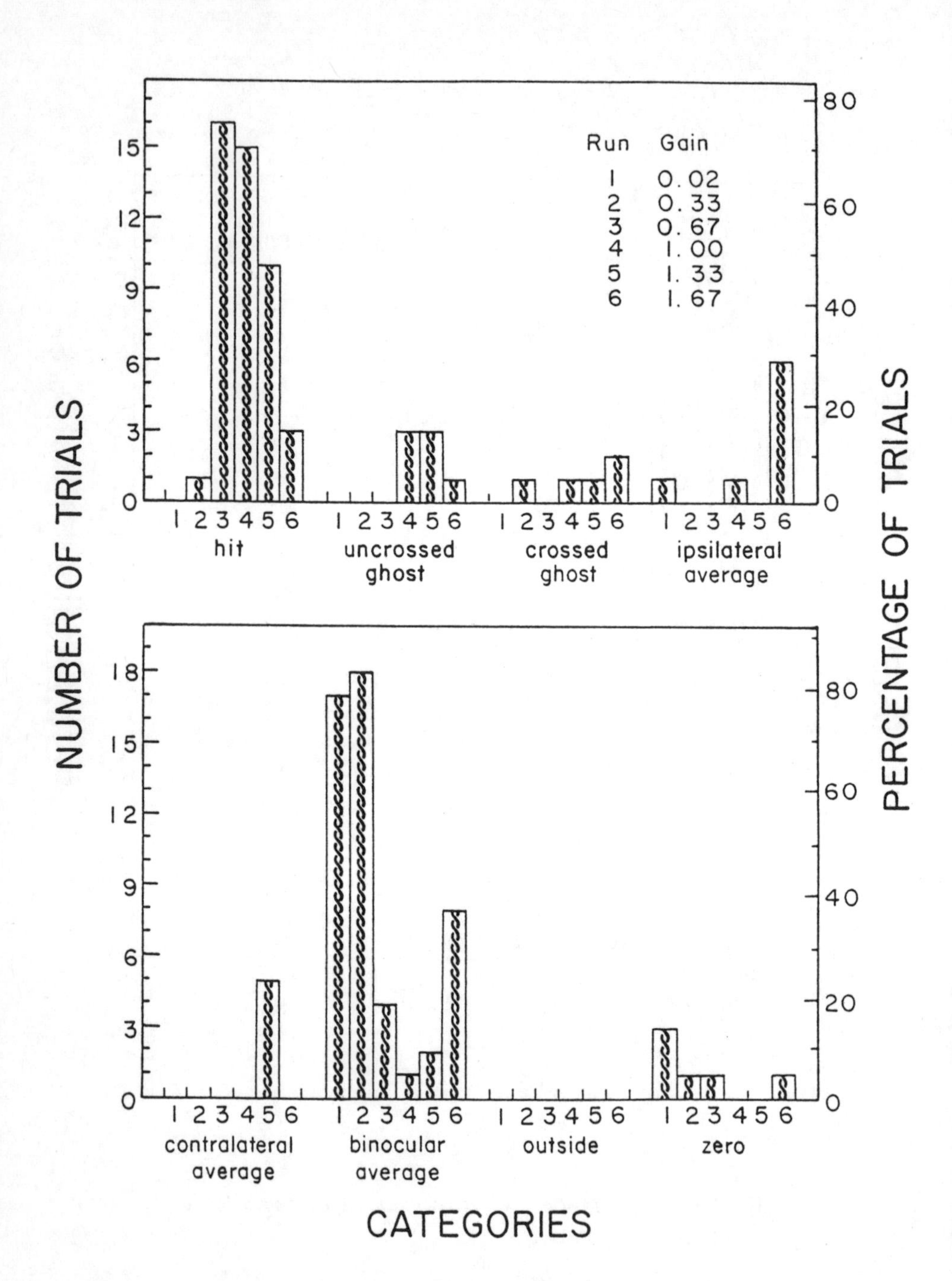

FIGURE 4.11. Model Performance vs. Binocular Gain

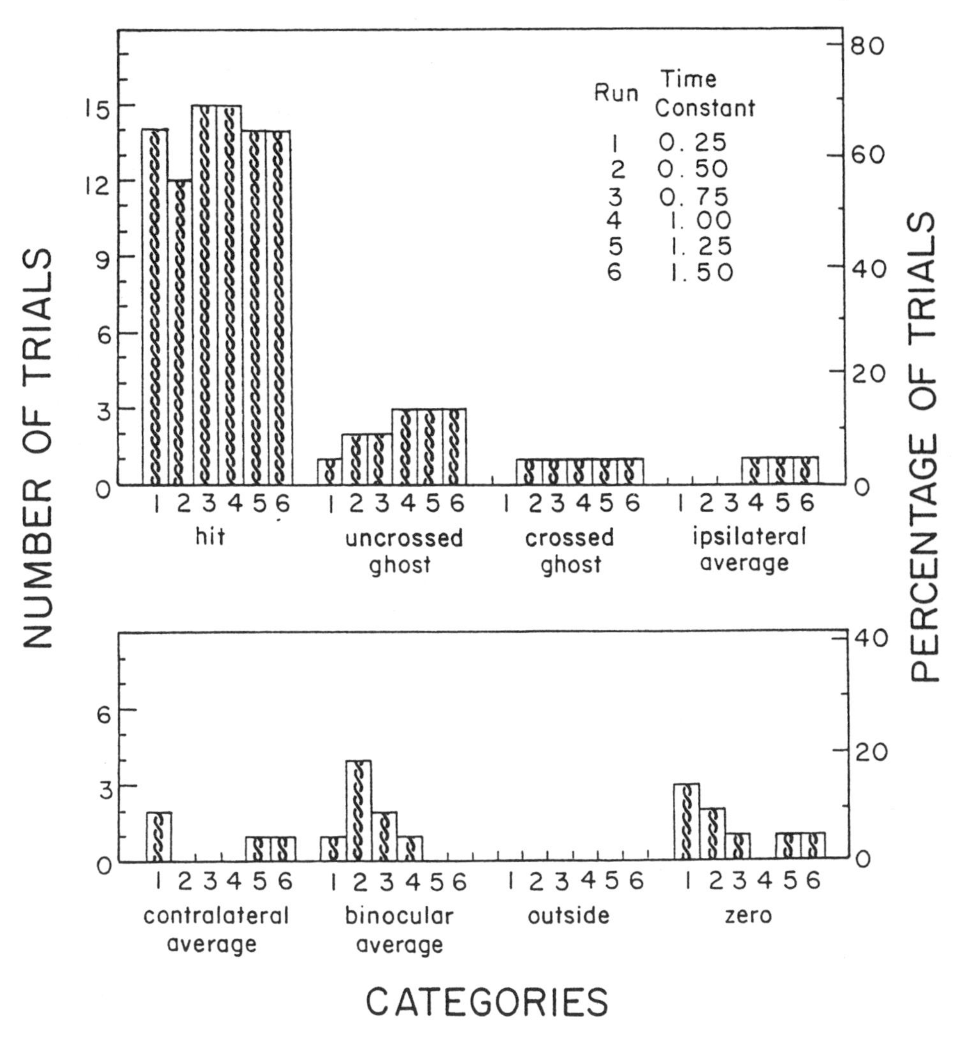

FIGURE 4.12. Model Performance vs. Accommodation Time Constant

a broader region. This also accounts for the fact that with high binocular gain there are a significant number of cases of both *ipsilateral-average* and *contralateral-average* errors. In both of these cases, one of the eyes selects a prey object while the other selects an average angular location.

The time constant T_a of the accommodation controller had a less dramatic but noticeable effect upon model performance. Fig. 4.12 shows the results of experiments investigating the effect of varying this time constant. Run 4 was done with the time constant set at its nominal value. Although the value of the time constant does not have a large effect on the total number of *hits* versus incorrect depth estimates, there is an effect upon the types of errors made. With a small time constant (25-50% of nominal) most of the errors are *zero* or *average* errors. However, as the time constant is increased to 150% nominal, the number of *uncrossed-ghost* errors increases.

The high number of *zero* and *average* errors at low time constants is due to the fact that the lenses are accommodated to a particular depth before the prey selectors have converged. Thus accommodation depth tends to be inaccurate. The resulting defocus has an effect similar to that obtained by lowering the binocular gain.

4.2.3 The effect of lenses on two-prey experiments

Collett and Udin [1983] were able to produce snapping errors in both normal and lesioned toads by placing concave lenses in front of their eyes. We were able to partially reproduce this effect with our model. Fig. 4.13 shows model performance on the symmetrical test suite when simulated concave lenses of various strengths were interposed between the eyes and the test scene. As lens strength was increased, the number of errors made by the model also increased. The errors fell mostly into the *crossed-ghost* and *binocular-average* categories. Collett and Udin's description indicates that actual toads with lenses affixed in front of their eyes exhibit *binocular-average* errors as well as a marked tendency to avoid snapping.

4.3 Discussion

We have demonstrated a model of depth perception whose development was guided by consideration of the prey localization process in frogs and toads. The model does not attempt to build a global depth map of a visual scene but rather locates a single point in space. It utilizes depth from binocular matching to guide lens accommodation that, in turn, assists in disambiguating binocular matching.

4.3.1 A possible neural realization of the model

The structure of the model was influenced strongly by our study of the structure and function of the visual system of frogs and toads. Our proposal for the way in which the model's structure might be implemented in this system is shown in Fig. 4.14. In this figure the two sides of the visual system are designated right (subscript R) and left (subscript L) based on functional rather than anatomical considerations. An anatomical representation would show crossed projections from each retina to contralateral tectum and both crossed and uncrossed projections from tectum to the motor area. The elements of the visual system of frogs and toads that have analogs in the model are the lens and its accommodation mechanism, the retina, the optic tectum, the nucleus isthmi, and the motor area.

Although there is little information available about the accommodation

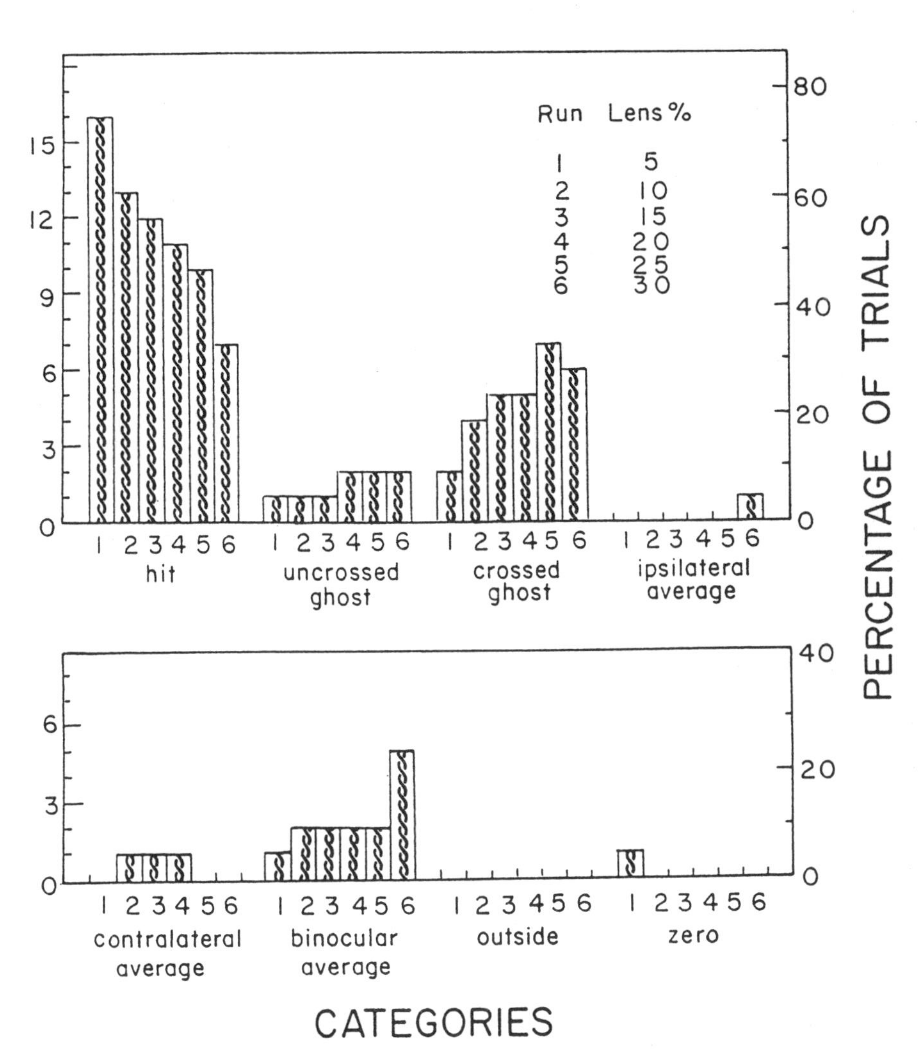

FIGURE 4.13. Model Performance with Interposed Lenses

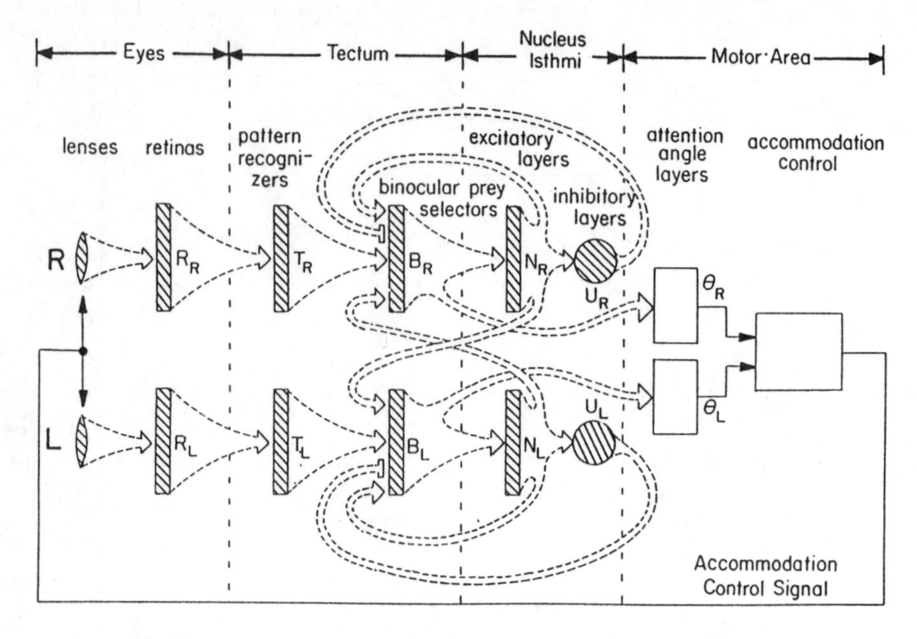

FIGURE 4.14. A Possible Neural Realization of the Model

mechanism in frogs and toads [Grüsser and Grüsser-Cornehls, 1976], it is known that the two protractor lentis muscles move the lens along the optical axis to effect a change in focal length [Jordan et al., 1980]. Since adequate physiological data are lacking, we turn to the lens/prism studies of Collett [1977] for evidence that lens accommodation is sufficient to provide depth cues adequate for accurate prey snapping. In relating the lens accommodation system to the model, we made the assumption that the lenses are coupled so that they both accommodate to the same depth. This assumption requires experimental verification.

There is indirect evidence to support the assumption that the strength of retinal ganglion-cell signals is a function of the degree to which the retinal image is correctly focused. The frog retina has been shown to contain at least four types of pattern selective units that project to tectum [Lettvin et al., 1959]. The receptive fields of retinal type-2 units, which provide the bulk of retinal input to tectum [an derHeiden and Roth, 1983], are small (4° to 8°) and consist of an excitatory center with an inhibitory surround. This kind of unit should be maximally sensitive to small, sharply focused visual stimuli.

Each tectal hemisphere receives retinotopically organized stimulation from the contralateral eye. The construction and location of the eyes is such that there is a significant region of binocular overlap in the frontal superior visual field, and both tectal hemispheres receive retinal information from this region [Fite and Scalia, 1976; Grobstein et al., 1980].

The function of pattern recognition within tectum is well known and

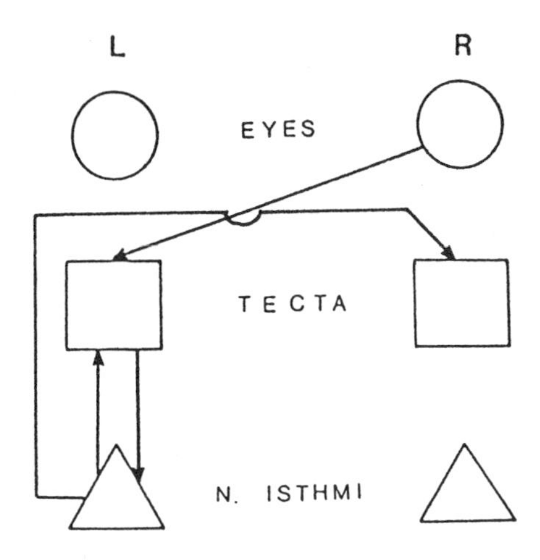

FIGURE 4.15. Block Diagram of Isthmio-Tectal Projections

has been the subject of extensive study by ethologists [Ewert, 1976] and modelers [Ewert and vonSeelen, 1974; Lara et al., 1982; an derHeiden and Roth, 1983; Cervantes et al., 1983]. Tectal units have been identified that are sensitive to object elongation along the axis of movement, direction of movement, figure-ground contrast, etc. In our model we assume that these feature detectors are able to isolate prey stimuli from the visual background.

The source of the binocular information necessary to enforce agreement between the two sides of the brain could be the nucleus isthmi. Each nucleus isthmi receives its input solely from the ipsilateral tectal lobe [Gruberg and Lettvin, 1980] and sends efferent projections to both the ipsilateral and contralateral tectal lobes. The relationship between tectum and nucleus isthmi is shown in block diagram form in Fig. 4.15. The tecto-isthmal and the isthmio-tectal projections are topographically organized so that the ipsilateral tecto-isthmio-tectal pathway maps a locus on tectum back to itself. The crossed connections map the binocular region of one tectal lobe to the binocular region of the opposite tectal lobe. These maps are in register (with respect to a fixed horopter surface [Gaillard and Galand, 1980], and provide each tectal lobe with a source of binocular input. Crossed connections also exist between the monocular regions of the two tectal lobes. These map anatomically and not visuotopically similar loci to each other [Grobstein and Comer, 1983]. Lesions to either the nucleus isthmi or its crossed efferent pathway abolish ipsilateral tectal stimulation [Glasser and Ingle, 1978; Grobstein et al., 1978]. Thus it is well established that the nucleus isthmi is the primary way-station in an intertectal binocular relay.

We assume that tectal binocular units receive three distinct excitatory inputs. Two of these inputs come from the two sides of the tectum via relays in nucleus isthmi with a third input from pattern recognition cells

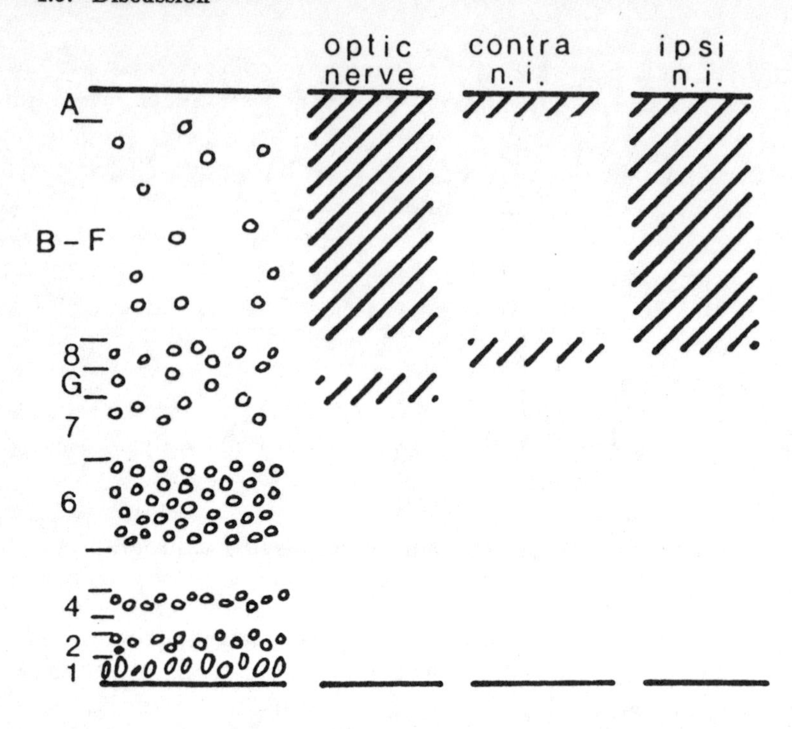

FIGURE 4.16. Distribution of Fiber Terminals in Superficial Tectum
Layers of tectum are shown from the most superficial layer (A) to the ependymal layer (1). Shaded areas show the distribution of terminals of the optic nerve, the contralateral nucleus isthmi, and the ipsilateral nucleus isthmi. Reprinted by permission from Gruberg and Lettvin [1980].

coming entirely from within tectum. This assumption is not based only upon the carefully preserved topography in the tecto-isthmio-tectal connections. The fibers of the crossed and uncrossed isthmio-tectal pathways terminate in distinct tectal laminae, as shown in Fig. 4.16. Fibers from both nuclei terminate in the vicinity of layer 8 where there are few optic-nerve terminals. The fibers from the nucleus isthmi carry information originating only from tectal efferents and thus may be assumed to result from visual events *significant* to the animal. In particular, prey activity in the visual field would presumably excite a high degree of activity. Signals from both nuclei experience similar delays between the original optic input and their arrival in layer 8 [Udin, personal communication]. Thus they are properly aligned both topographically and temporally, probably carry only visually significant information, and are free from the presence of direct optic-nerve input. This configuration seems particularly well suited for providing the binocular cross-coupling required by the model's prey selectors. We assume that the tectal binocular units receive this input across a broad receptive field [Fite, 1969; Raybourn, 1975].

Fig. 4.16 also shows that just below layer 8, in layer G ,is a region innervated by the optic nerve but without input from the nucleus isthmi. This is what would be required to provide the input to the pattern recognizers, since to be effective they must have a fairly pure direct input from the retinal ganglion cells. We assume that tectal binocular cells receive input from tectal pattern recognition units across a narrow receptive field. We picture the binocular cells as responding optimally when they are receiving input from all three sources: ipsilateral nucleus isthmi, contralateral nucleus isthmi, and pattern recognizers. The inputs from nucleus isthmi weight cells in binocular correspondence, and the inputs from the pattern recognizers assure angular acuity.

Besides providing excitatory feedback to the tectal prey selectors, the nucleus isthmi may also provide the global inhibitory signal required by the prey selectors. There is some evidence that, in addition to the provision of binocular relay by the isthmio-tectal projections, there is a second isthmio-tectal system providing tectal inhibition. Glasser and Ingle [1978] noticed that frogs with large lesions to nucleus isthmi show an elevated level of spontaneous tectal activity. Grobstein et al. [1978] report that there seem to be two distinct crossed projections from the nucleus isthmi. The projection involved in binocularity arises from the ventral and medial rim of the nucleus and projects to rostral tectum. A second projection arises from the medullary region of the nucleus and projects to the entire tectum. This projection may be inhibitory. For instance, although there is an anatomically well defined projection to monocular tectum, there have been no successful recordings made of activity in monocular tectum elicited by visual stimulation of the corresponding monocular region of the ipsilateral eye. Although this evidence is not conclusive, we make the assumption that the projection from the medullary area produces a global inhibition in tectum proportional to the total level of activity in the tectum. This assumption is also attractive from an anatomical point of view, since each of these nuclei takes as its only input a topographic map of tectal activity, and concentrates this map into a relatively small region. Thus these nuclei have the structure and neural pathways necessary to provide a signal to tectum that is modulated by the average (or total) firing rate across the entire tectal surface.

Finally, tectal efferents, pictured here as coming from broad-field binocular cells, descend into the motor area where they initiate actions of the animal [Ingle, 1983; Grobstein et al., 1983]. We assume that one such motor activity is adjustment of the accommodative state of the lens and that this motor activity precedes any overt action in prey catching.

4.3.2 Suitability of the Nucleus Isthmi as a Tecto-Tectal Relay for Depth Perception

If cross-tectal projections via the nuclei isthmi are to provide the binocular data for prey selection, they must be sufficiently dense to account for the accuracy in depth determination shown by both frogs and toads. Each nucleus isthmi contains about 8000 cells [Gruberg and Udin, 1978; Wang et al., 1981], and each of the crossed and uncrossed isthmio-tectal pathways contains about 4000 fibers [Gruberg, personal communication]. In order to estimate the error in angular disparity to be expected from this number of fibers, we make the following assumptions: (1) the fibers innervate the neural representation of a 180° (horizontal) by 90° (vertical) visuotopically organized section of a cone (Gaillard and Galand [1980] describe the horopter in frogs as being cone-like); (2) the fibers are organized into evenly spaced concentric semicircles, each representing a fixed angular elevation in visual space; and (3) the number of fibers allocated to a particular elevation is proportional to the circumference of the semicircle representing that elevation.

A simple calculation, based upon these assumptions, shows that if the ground-plane semicircle is allocated 180 fibers (one per horizontal degree) then nearly 45 such rings can be constructed from 4000 fibers. Thus the horizontal angular resolution on the ground-plane will be $\pm 1/2°$. The vertical resolution will be $\pm 1°$. The worst-case horizontal disparity error on the ground plane, combining two such representations, will be twice the maximum error in horizontal acuity or $\pm 1°$. Fig. 4.17a shows the approximate relationship between angular disparity and depth for objects located in the center of the binocular field. This curve was calculated using simple pin-hole optics and an assumed eye separation of about 3 cm. Fig. 4.17b, computed from the slope of the curve in Fig. 4.17a, shows the expected maximum depth resolution error as a function of depth. From this curve we see that the effect on depth estimation of the cumulative 1° error in angular disparity would be less than 0.2 cm for objects at 5 cm from the animal. At 10 cm the depth error increases to 0.6 cm and at 15 cm it is just greater than 1 cm. These figures correlate well with animal behavior. For instance, the prey overshoot observed by Collett [1977] was between 0 and 2 cm for prey distances out to 15 cm.

4.3.3 Evaluation of the Model

Although the model can be directly related to neurophysiological and neuroanatomical data, the algorithm derived from the model is interesting in its own right. In particular, it may be useful as a mechanism for depth perception in robots when the problem is the detection and localization of a particular object. Simulation results demonstrate that this algorithm is able to rapidly locate a single object placed anywhere in the binocular visual field, and to select and localize the nearer of two identical objects when they are presented at different depths. Further, the algorithm is successful 75% of the time in selecting and localizing one of two objects over a large range of perfectly symmetrical presentations.

In toads, Collett [1977] noted a 94% effect from binocular cues and a

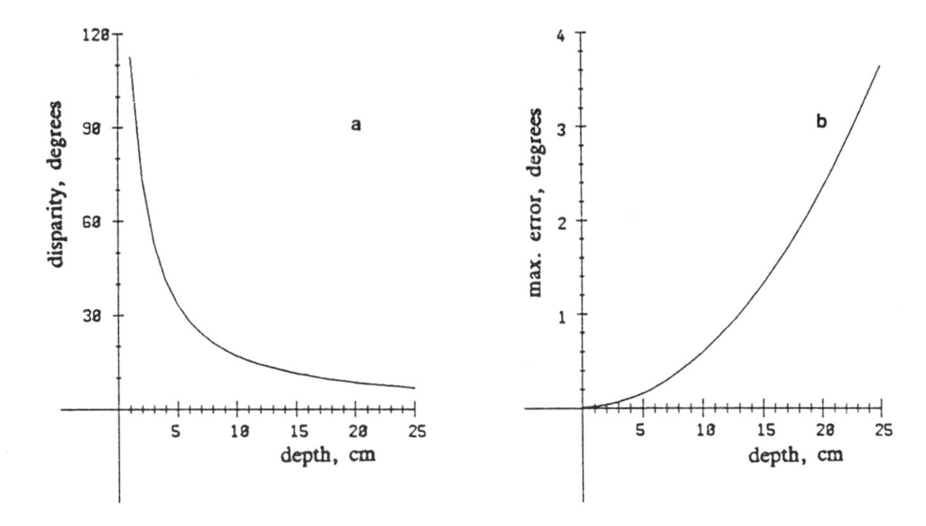

FIGURE 4.17. Angular Disparity vs. Distance from Viewer

6% contribution from monocular cues. However, when single-prey presentations are made to our model, lenses have no effect upon depth resolution, whereas prisms cause the expected shift in depth estimation. This is because the model contains no mechanism to allow for any direct monocular contribution to depth estimation. It computes its depth estimate solely from binocular parallax. Thus an obvious extension of the model would be to provide a secondary depth estimator from image focus cues. Since frogs and toads are successful at monocular prey-localization, it is likely that such a system exists. However, our model demonstrates that optimizing image focus is not the only possible mechanism for controlling lens accommodation. In fact, it is likely that in frogs and toads both image focus and binocularity play a cooperative role in adjusting the lens.

Our experiments with simulated concave lenses and symmetrically presented prey showed that the model is prone to making both *crossed-ghost* and *binocular-average* errors under these conditions. On the other hand, toads seem to make only *binocular-average* errors under identical circumstances [Collett and Udin, 1983]. If the model employed both binocular and monocular cues to adjust the lens and estimate depth, then the number of *crossed-ghost* errors would diminish and the frequency of *average* errors would increase. In other words, the depth estimate from binocular mismatch would be offset by the depth estimate from lens focusing. Toads rapidly learn to avoid snapping altogether under these conditions [Collett and Udin, 1983]. This fact also argues for a dual-cued system. The discrepancy between the depth estimates provided by these two cue sources would be sufficient to "abort" the toad's normal prey-catching activity.

Of the categories of depth estimate shown in Fig. 4.6, both the *uncrossed-*

ghost and *ipsilateral-average* categories can provide depth estimates that are behind the eyes. A more sophisticated accommodation controller could be designed to improve performance based on this fact. Such a controller would use the depth setpoint, calculated from disparity in the prey selectors, only if its value were reasonable.

5

Towards a Complete Model

ABSTRACT In this chapter we compare and contrast our two models, first with each other and then with data from biological experiments. Our analysis suggests that frogs and toads do not use a single means for determining depth but, instead, use two different processes, one for determining depth of stationary environmental objects and the other for determining depth of prey. We present an accommodation control system that meets the needs of both of these processes, and show how some of the discrepancies between animal experiments and our *prey localization model* would be eliminated by the nonlinearities inherent in this control. The chapter concludes with an evaluation of the depth scale used in the two models.

5.1 The Cue Interaction and Prey Localization Models

5.1.1 Unities

The *cue interaction model* shares several features with the *prey localization model*. These commonalities are not accidental, but rather point to certain underlying charactersitics of the depth resolution problem faced by frogs and toads.

The most important and most obvious similarity is that both models use lens accommodation to help disambiguate the binocular matching process. Further, it is depth from binocularity that determines the final depth output from each of the models. Obtaining depth from binocular matching is ambiguous without auxiliary depth cues, and such cues are available from lens accommodation in both frogs and toads. The models demonstrate two different ways in which image focus can be used to disambiguate the cues from binocular matching. The depth maps computed by the *cue interaction model* are dominated by binocular cues, and in the *prey localization model* the final lens accommodative state is determined by the binocular disparity between selected points. That the final depth estimate should come from binocularity and not from lens accommodation is based upon both experimental findings [Collett, 1977] and theoretical considerations [Collett and Harkness, 1982].

The models are also similar in their computational schemes. They are both based upon competition and cooperation. Arbib [1981] makes a convincing argument that cooperative computation is the style of processing in the brain. Amari and Arbib [1977] provide a short review of some cooperative models and show how these models can be unified by a consistent mathematical formalism. Their presentation is particularly applicable to the problem of depth perception. Marr and T. Poggio [1979], on the other hand, argue that cooperative computation is not necessarily style of computation used for the depth perception (for a counter argument see Frisby and Mayhew [1980]). However, since toads can orient to the position of a prey object even after it is no longer visible [Collett, 1982], whatever computational style *is* used must provide some form of short-term memory. Due to their inherent hysteresis, cooperative-competitive schemes naturally provide a stable interpretation of sensory input.

A more subtle similarity between the models is that, while both models depend on binocularity for determining depth, they achieve angular acuity by sampling the visual input at a point before any binocular interaction takes place. In the *cue interaction model*, the retinal position of a point of visual input is used to index a depth map to determine the spatial position of the point. Retinal position is required since the depth maps produced by the model have good depth acuity but poor angular acuity. In the *prey localization*, monocularly driven pattern recognizers provide a precise angular position signal to the prey selectors. The selectors' binocular inputs are more broadly spread in the angular direction. These precise angular signals, which both models use, do more than provide directional acuity. Their source is close to the retinal input so they are relatively free from hysteresis. Thus, although neither model was designed to address the problem of changing visual stimuli, they both contain a mechanism that would allow them to be responsive to change.

5.1.2 Diversities

Although the two depth models have several commonalities, they are more easily contrasted than compared. All of the important differences between the models stem from the fact that depth perception is viewed as an information-gathering process in the *cue interaction model*, whereas in the *prey localization model* it is represented as an action-oriented information-seeking process.

In the *cue interaction model* there is no implication that any action is taken based upon the data present in the visual input. This model represents the information flow in the nervous system as being unidirectional. Any action based upon the output of the model is presumed to be taken by motor centers downstream of the depth-mapping process. If, for example, a new object were introduced into the visual field, the response of the *cue interaction model* would be merely to adjust its depth map.

The *prey localization model*, on the other hand, utilizes an active coupling between sensory and motor processes. It represents the levels of the visual system as being part of an information-seeking feedback loop. Information flow is circular, completing a cycle of action and perception that is specialized to the capture of prey. In this model, a prey object introduced into the visual field would either be ignored, if the new input were less salient than that from an object already selected, or would itself become the selected object and cause an adjustment of the accommodative state of the lenses. This adjustment would serve both to verify the binocular depth estimate of the object and to improve its visual acuity. Thus, rather than passively gathering data, this model pictures the visual system as commiting its resources to the examination of an object.

Since the *cue interaction model* is structured around the production of depth maps and the *prey localization model* is designed only to localize a single point, the two models require very different lens accommodation mechanisms. The *cue interaction model* requires an accommodation system that is able to continuously provide a depth likelihood (or image-focus) measure at every retinal position, and for all states of lens accommodation. An accommodation controller meeting this requirement would have to combine some form of accommodation scanning mechanism with a short-term memory, or map, recording image intensity at each lens setting. On the other hand, the accommodation system used by the *prey localization model* simply computes a single depth from binocular disparity and then adjusts the lens to bring that depth into clear focus.

5.1.3 A SYNTHESIS

Even though the two models represent very different ways in which both binocularity and lens accommodation can be used to determine depth, it is not necessary that we choose one of these mechanisms over the other. Instead, we will argue that the differences between these models specialize them for two equally important roles, and that mechanisms similar to those used in both models are essential to support the full range of activities exhibited by frogs and toads. Our conclusion is that multiple depth-resolving systems may coexist in a single animal.

To navigate successfully through their environment, frogs and toads would need a reasonably complete "picture" of the spatial configuration of surfaces in that environment. Toads are able to perceive simultaneously the positions of several barriers, gaps, and chasms. During prey approach behavior, they coordinate their activity based upon rules that take into account the complete spatial configuration confronting them [Collett, 1982; Arbib and House, 1983; Lara et al., 1983]. The *cue interaction model* builds the kind of spatial depth map needed to support this observed behavior. It is also best suited to the kind of visual input presented by natural barriers. This input, from grasses, twigs, rocks, etc., would more closely resemble a random-dot stereogram than would the very sparse input from prey. The algorithm underlying the *cue interaction model* [Dev, 1975; Marr and Poggio,

1976; Amari and Arbib, 1977] was originally designed to explain perception of depth in random-dot stereograms and is especially effective with this kind of data. Since the *prey localization model* locates only a single point, it is ill-suited for determining the depth of barriers.

For catching prey the advantage lies not in determining the positions of all prey-like objects but in chosing a single prey object and then directing all sensory and motor activity towards its capture. The *prey localization model* is best suited for this form of depth resolution. It selects a single prey object and then maximizes the quality of the image of that prey ojbect by adjusting the lens focus. It even has a good source for a cue to initiate a strike at the prey; when the lens accommodative state matches the depth estimate from binocularity, there is a high degree of certainty that a correct depth measurement has been reached and a strike should be initiated. The more complex depth-mapping machinery of the *cue interaction model* is neither necessary nor efficient for prey capture.

Our hypothesis, that prey and barrier depth are determined by two different processes, is also supported by anatomical, physiological, and behavioral evidence.

A large number of experiments indicate that prey acquisition in frogs and toads is tectally mediated, whereas barrier avoidance is mediated by the thalamus. Extensive studies have been made of tectal units sensitive to small moving prey-like objects [Ewert, 1976; Ewert, 1982]. Functional units specialized for stationary edge detection have been discovered in thalamus [Ewert, 1971; Brown and Marks, 1977; Ingle, 1980]. The blue-sensitive units discovered by Muntz [1962] also respond to stationary input. Frogs with bilateral tectal ablations do not exhibit prey-catching behavior but do retain the ability to avoid barriers [Ingle, 1977]. Motor pathways reflect the same organization. Severing the tecto-fugal pathways disrupts orientation and snapping behavior, whereas severing the pretecto-fugal pathways disrupts sidestepping to move around barriers [Grobstein et al., 1983; Ingle, 1983]. There is partial evidence that this differentiation of function extends to depth perception; atectal frogs are able to discriminate the depth of barrier surfaces [Ingle, 1982].

The functional differentiation between tectum and thalamus is matched by anatomical differences that also support our hypothesis. The computation done by the *cue interaction model* is most easily supported by a structure, like the thalamus, that has overlapping representations of the visual fields of both eyes [Scalia and Fite, 1974]. The *prey localization model* is best suited to the organization exhibited by the tectum and nucleus isthmi, in which the tectum receives only monocular projections but the nucleus isthmi acts as a relay to pass information between tectal lobes [Glasser and Ingle, 1978].

Our proposal for the distribution of depth mapping and prey localization processes in the brains of frogs and toads is shown in Fig. 5.1. The thalamus is shown receiving a binocular projection from the eyes as well as signals in-

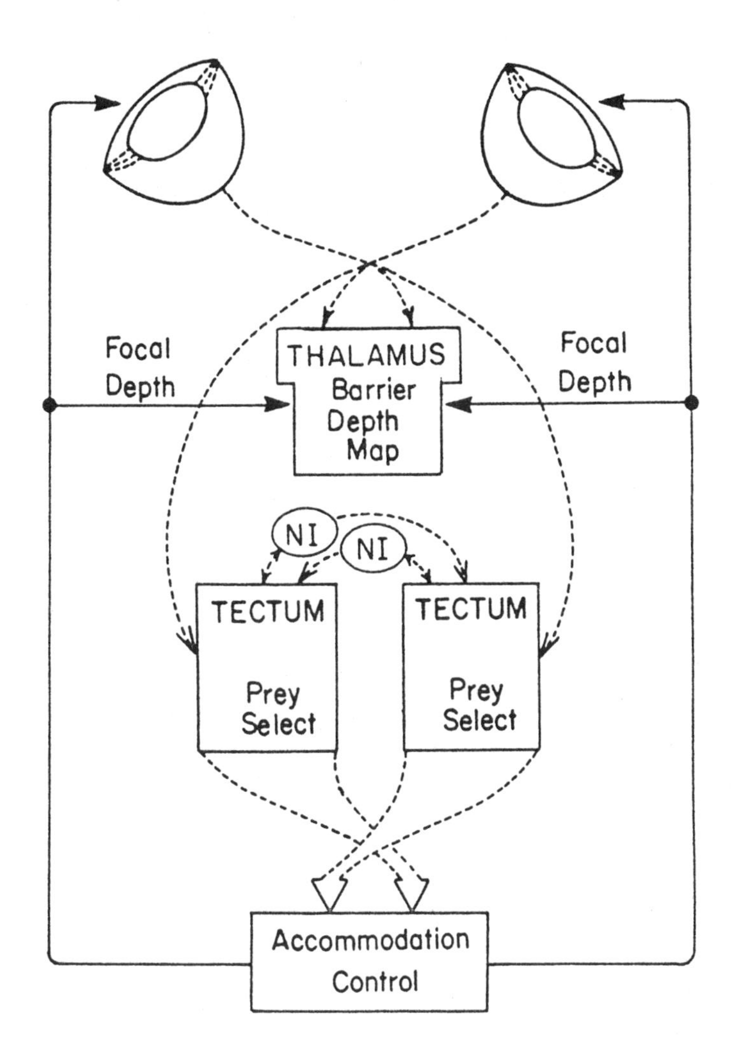

FIGURE 5.1. An Anatomical Distribution of the Two Depth Processes
Eyes are represented at the top of the figure, and internal brain regions are indicated by labeled rectangles. Dashed arrows depict information inflow, and solid lines represent outflow.

dicating the lens setting being called for by the accommodation controller. (Rubinson and Colman [1972] report that the nucleus posterolateralis in the thalamus of frogs (*Rana pipiens*) receives an afferent projection from the midbrain tegmentum. However, its physiological function is unknown.) These inputs would provide the thalamus with the information required to make depth inferences based on both binocularity and lens accommodation. A process similar to that used in the *cue interaction model* could use these inferences to build depth maps. The tectum is pictured as receiving monocular projections, but also has relay pathways through the nucleus isthmi that allow information interchange between the two tectal lobes. As in the *prey localization model*, projections from tectal prey selectors would provide the accommodation controller with information for computing a binocularly-derived depth setpoint.

5.2 An Extended Model of Accommodation Control

The hypothesis that these two different depth perception processes coexist in frogs and toads requires that we consider a more sophisticated model of accommodation control than was posited in either of the two previous models. If we assume that accommodation is used to provide monocular depth cues that assist in determining the depth of both barrier and prey objects, then the accommodation control system must be capable of at least two modes of operation.

The *cue interaction model* requires that the brain be able to construct a map of image intensity indexed by both retinal position and lens accommodative state. In order to construct this type of map, the lens would have to be scanned, so that over a period of time a complete map would be developed. This mode of accommodative control is appropriate for the depth mapping of stationary barrier-like objects, but is unlikely to be effective for the mapping of moving prey-like stimuli.

The *prey localization model* requires that the accommodation controller be able to lock on to a prey stimulus quickly, setting the lens accommodative state to match the true depth-in-space of the prey. This mode of accommodative control is appropriate for isolated stimuli, and its speed would allow it to be responsive to a moving stimulus.

Even if the accommodation controller were to have the two modes of operation described above, it would still not be capable of representing some of the more subtle depth-related behaviors exhibited by frogs and toads. Toads judge the depth of prey located in their frontal binocular field using mostly binocular depth cues. Nevertheless, their depth estimates do include a small (6%) effect from accommodation. When concave lenses are placed in front of their eyes or when drugs are used to disrupt accommodation, binocular toads make small errors in depth estimation [Collett, 1977; Jordan et al., 1980]. The *prey localization model* does not replicate these errors since it does not use monocular cues in its estimation of depth. The question remains: How are accommodation and binocularity integrated for prey catching?

The following findings from experiments by Collett and Udin [1983] are pertinent to this question. Over a range of experimental conditions, estimates of the depth of a single prey object made by toads with lesions to the nucleus isthmi are very similar to those made by intact toads. But there are important differences between the estimates made by lesioned and unlesioned animals when two prey objects are presented simultaneously. In this case, the lesioned toads will sometimes snap at a "ghost" prey located between the two true prey objects, and near the point in space corresponding to the position predicted from binocular mismatch (see the *crossed-ghost* error of Fig. 4.6b). However, the error they make is not *exactly* that predicted by binocular mismatch; snaps made toward "ghost" prey consistently overshoot the depth expected from a true binocular mismatch error. When concave lenses are mounted in front of their eyes, the lesioned toads see "ghosts" at the same distance as they do without the lenses. However, when the lesioned toads view two prey objects through lenses and select a *real* prey object, their snaps fall short of the true target but are considerably closer to the prey than their snaps at "ghosts." This result is seen only in nucleus-isthmi lesioned animals, and is similar to results obtained in earlier experiments with monocular toads [Collett, 1977].

Our *prey localization model* predicts "ghost" snapping when the nucleus isthmi is lesioned but it does not predict the more subtle effects apparently due to the interaction between binocular and accommodation cues. To realize these effects, we propose three additional extensions to the accommodation controller. First, we introduce a deadband into the controller so that control action will be highly responsive to large errors between the binocular setpoint and the current lens setting but will be unresponsive to this error when it is small. Second, we introduce a control signal derived from image focus cues that will act to improve the focus of the retinal image. Third, the use of image focus cues requires that the dynamics of the controller be replicated so that there is a separate control path for each lens. If only binocular cues are used by the controller, both lenses will receive the same depth setpoint, so there need be only one control path. However, image focus cues will be different for each eye, requiring two separate control paths. With these additions in place, the controller could be adjusted to use binocularity to achieve a rapid "approximate" lens adjustment, and focus cues to make a further "fine" adjustment.

The fully extended version of the accommodation controller is shown in Fig. 5.2. As in the accommodation controller of the *prey localization model*, the vector summation boxes receive input from the tectal prey selectors. In addition to providing retinal positions θ_L and θ_R, denoting the centers

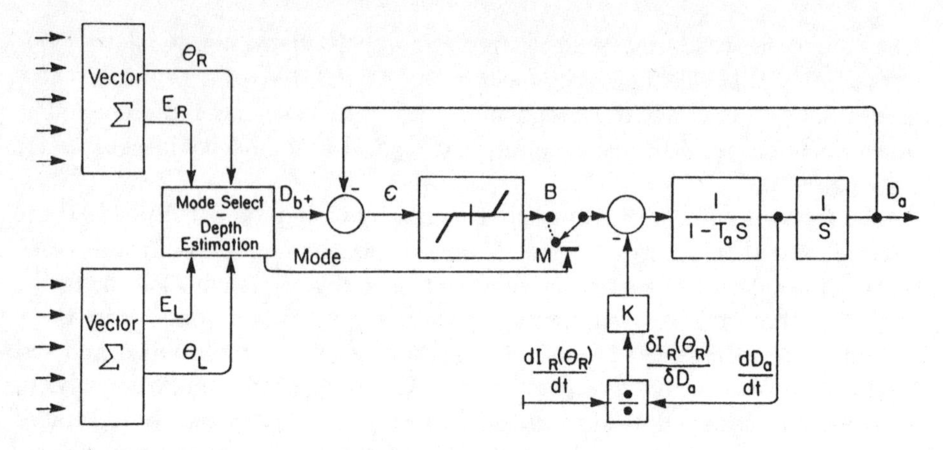

FIGURE 5.2. Extended Accommodation Controller

of gravity of excitation in the two prey selectors, the summers also provide level signals E_L and E_R, which measure the intensity of excitation in the prey selectors. The "mode-select" box has two outputs: D_b, the binocular depth setpoint computed from the difference (or disparity) between positions θ_L and θ_R, and a mode selection signal that switches the controller between monocular M, and binocular B modes. The error signal ϵ, the difference between the depth setpoint D_b and the current accommodative setting D_a, is shown passing through a deadband that effectively decouples binocular control when lens accommodative state and the binocular setpoint are nearly identical. The controller is also augmented by a circuit that measures the change in image intensity I with respect to change in lens accommodative state $\partial I_R(\theta_R)/\partial D_a$. This circuit provides positive feedback through a gain K, so that when the system is decoupled from binocular control (either because of the mode switch or the deadband) the lens will be adjusted in a "hill-climbing" fashion to improve image focus.

There are various ways in which mode selection could be done in the controller. The simplest would be to compare excitation levels E_L and E_R with small thresholds, with monocular mode selected if either level is below threshold and binocular mode selected if both are above threshold. In addition to thresholding, a more sophisticated mode selector could apply consistency constraints. For instance, a requirement could be that both excitation levels have similar values. Another could be that the depth setpoint D_b computed from the difference between positions θ_L and θ_R correspond with a location in the frontal binocular field of the animal.

In the *prey localization model* it is assumed that the two lenses are coupled so that they are always adjusted to the same accommodative state. In this extended version of the controller this constraint is relaxed. The dynamics of the controller are replicated, so that there is one control path for each lens. When accommodation is under binocular control, both paths receive the same setpoint D_b so that their action is coupled. When under monocular or image focusing control, each control path acts to maximize independently the clarity of focus of the image on the side that it is controlling.

In order to support its monocular focusing mode, the controller would require two very different kinds of input from higher brain centers. First, it would require inputs that are not heavily processed, and therefore closely mirror the changing pattern of retinal excitation. These inputs would provide the controller with the sensitivity to retinal activity required to detect small changes in the quality of image focus. Projections to the controller directly from thalamic and tectal pattern sensitive cells could provide this sort of input. The controller also would require inputs that would govern the location of specific local regions on the retinas that should be monitored for determining the quality of focus. The tectal prey selectors, already included in the accommodation controller of the *prey localization model*, could provide this input for prey stimuli. For barrier stimuli, signals indicating the most salient barrier edges would be sufficient.

In frogs and toads, there are tectal and thalamic efferent pathways that could supply these two kinds of input. Tectum and thalamus each have two major efferent pathways to the brainstem. In both cases, one of these pathways has its origin close to the terminal arborizations of the optic nerve, and the other pathway has its origin farther from these terminals [Ingle, 1983; Grobstein et al., 1983] (see Fig. 2.9)

The augmented accommodation controller, described above, has a structure that would allow it to replicate the data from behavioral and lesion experiments. First, the controller could be tuned so that its response would be dominated by binocular cues when they were available, at the same time providing for a small contribution from monocular cues. If the gain K on the image focus feedback is kept small, and the controller is in binocular mode, then this feedback will have only a negligible effect as long as the error signal is large enough to be outside of the deadband. But when the error is reduced so that it is within the deadband, the focus signal will have its full effect. However, the range of this effect is limited by the deadband. If the focus mechanism adjusts accommodation enough to move the error outside the deadband, binocular control will be reinstated and will return the error to within the deadband. If we assume that the depth estimate used by the animal for snapping is the controller's output signal D_a, then the presence of concave lenses in front of the eyes would cause undershooting, but only to the extent permitted by the width of the deadband. If the *crossed-ghost* error were being made by the binocular system, as in nucleus-isthmi lesioned toads, then the focusing system would tend to move the lens to correct this error but, again, this adjustment could only go as far as the deadband and binocular control would allow. The resulting depth estimate would be somewhat farther away than the estimate that would have been made by binocularity alone. Concave lenses placed in front of the eyes would have no effect on the estimation of the depth of "ghosts," since the misfocus due to binocular error would be much greater than any effect from the lenses.

The effect of lenses on the estimation of the depth of real (as opposed to "ghost") targets in lesioned toads could be explained by the separation of the accommodation control paths for the two eyes. If the strength of the concave lenses were great enough, then the defocusing effect might be enough so that one side of the visual system, presumably the side ipsilateral to the selected prey, would have the object in better focus than the other side. The difference in strength between the signals from the two prey selectors could be interpreted by the controller as a binocular mismatch. The binocular decoupling due to the lesions to the nucleus isthmi would reduce the strength of the input to the prey selectors, increasing the likelihood of this mismatch. Because of this mismatch, the controller would revert to monocular mode and the concave lenses would have their full depth-distorting effect, as if the animal were monocular.

We complete our description of the accommodation process by examining the effect that the addition of focus control would have when the controller is being driven in the accommodative scanning mode. In this mode, the presence of image focusing feedback would cause the lens to spend longer periods of time in the regions of its cycle where image intensity cues are greatest. Instead of passively scanning, the controller would be more highly responsive to regions of the image as they came into clear focus.

5.3 Experimental Verification of the Models' Depth Scale

There is experimental evidence providing a confirmation of the internal depth scale used in both of our models. The models use a disparity-based scale (see Appendix B), rather than a Cartesian scale, not only to represent depth estimates obtained from binocular disparity but also to represent depth estimates from lens accommodation. One result of using this scale is that depth estimates become progressively less acute with distance from the eyes. That there should be a reduction of depth resolution with increasing distance is intuitive. However, does a disparity-based depth measure provide the correct form for this nonlinearity?

Jordan et al. [1980] in their experiments using drugs to disrupt the action of the lens accommodation muscles in toads, compared the maximal snapping distance *MSD* of treated animals with that of untreated animals. *MSD* was defined as the distance at which toads first snap at prey when the prey is moved slowly toward them. If a perturbation of the visual system increases an animal's *MSD* then it will tend to undershoot; if the *MSD* is decreased, the animal overshoots. The average *MSD* for untreated binocular toads was 4.3 cm. When animals were treated with a relaxant of the

TABLE 5.1. MSD After Drug Treatment of Accommodation Muscles

BINOCULAR TOADS

	Untreated	*Relaxant*	*Contractor*
MSD, cm	4.3	5.0	3.7
ΔMSD, cm	– – –	0.7	–0.6
% full range (1.3 cm)	– – –	54	46
$\Theta, tan^{-1}(MSD/1.5)$	70.7°	73.3°	67.9°
$\Delta\Theta$	– – –	2.6°	–2.8°
$\Delta D, 2\Delta\Theta/90°$	– – –	0.058	–0.062
% full range (0.120)	– – –	48	52

MONOCULAR TOADS

	Untreated	*Relaxant*	*Contractor*
MSD, cm	4.6	7.2	3.2
ΔMSD, cm	– – –	2.6	–1.4
% full range (4.0 cm)	– – –	65	35
$\Theta, tan^{-1}(MSD/1.5)$	71.9°	78.2°	64.9°
$\Delta\Theta$	– – –	6.3°	–7.0°
$\Delta D, 2\Delta\Theta/90°$	– – –	0.140	–0.156
% full range (0.120)	– – –	47	53

Distance data from Jordan et al. [1980]

Calculations of angles were made assuming an interpupillary distance of 3.0 cm, and *MSD* measured along the midline. ΔD is change in disparity between the retinal projections of a point on the right and left retinas, with a change of 90° represented by a 1-unit change in disparity.

lens accommodation muscles, the average *MSD* increased to 5.0 cm, and when they were treated with a muscle contractor, *MSD* decreased to 3.7 cm. In monocular toads the difference was more extreme. Their average *MSD* when untreated was 4.6 cm, compared with 7.2 cm when treated with ther relaxant and 3.2 cm when treated with the contractor. Table 5.1 summarizes these data, and illustrates how they would translate onto a disparity scale. With disparity taken as the depth measure, the percentage contributions of relaxant and contractor to the full drug-induced change in depth estimate (relaxant depth – contractor depth) were nearly the same for both binocular and monocular toads. In both cases, the relaxant contributed about 47% of the change and the contractor contributed about 53%.

Table 5.1 also suggests reasonable upper and lower limits for the deadband that is applied to the error signal of the accommodation controller of Fig. 5.2. When binocular toads are subjected to full disruption of accommodation, they make depth estimation errors that are equivalent to a 2.8° error in retinal angle. Thus we may assume that binocularity will bring the lens to within ±2.8°, after which the final lens position, and depth estimate, is based upon maximizing image focus.

5.4 Discussion

We have argued that frogs and toads probably have at least two kinds of depth perception, one for deriving a depth map of stationary environmental objects, and the other appropriate for localizing and snapping at moving prey. If this dichotomy does in fact exist,then it is likely that the mapping process takes place in the thalamic region of the brain and the prey localization process takes place in the tectum.

Although we have suggested only two depth perception systems, there is no reason to suppose that our arguments could not be used to support even more systems. Along with certain anatomical, physiological, and behavioral evidence, our argument rests mainly on the grounds that perception is action-oriented. Different actions require different depth information. Thus what is necessary and efficient for one type of action may be either unnecessary or inefficient for another type of action. This argument may, of course, be extended to cover behaviors other than prey catching and barrier negotiation, such as mating, locating ponds, and journeying. The potential for an explosion in the number of depth perception processes required by a single animal suggests that theorists should attempt to determine a minimal set of such processes that could efficiently support the wide variety of behaviors requiring depth perception. We have suggested two such processes: (1) derivation of a complete environmental depth map, and (2) localization of a single point.

We have constructed a preliminary model for a lens accommodation controller that would be able to support both of these processes. This model contains a method for switching between a binocularly driven "prey selection" mode and a "barrier mapping" mode driven by image focus cues. The model is also configured to explain some of the effects seen in behavioral experiments with toads, when the correspondence between monocular and binocular depth cues is disrupted by placing concave lenses in front of an animal's eyes.

Behavioral data support the idea of a continuously developed depth map in thalamus. Atectal frogs are able to discriminate the depth of barrier surfaces during escape [Ingle, 1982] (see Fig. 2.10). Escape requires that an animal be able to respond immediately, and not after patiently constructing a depth map.

Our assumption that a depth map of stationary environmental objects is continually being built seems to contradict physiological evidence that frogs and toads show very little visually elicited neural activity when there is no motion in the visual field. However, this evidence was obtained from experiments with immobilized animals and is far from conclusive. Further, we believe that this depth mapping process goes on in the thalamus, where at least two kinds of neural units responsive to stationary visual input have been found. These thalamic units are the blue sensitive "on" units first reported by Muntz [1962] and the stationary edge detectors best described by Brown and Marks [1977]. The blue sensitive "on" units appear to be clustered in the region of the nucleus of Bellonci [Brown and Marks, 1977], which is one of several binocularly innervated sites in the thalamus [Scalia and Fite, 1974].

Most electrophysiological studies described in the literature involve the use of curare to paralyze the animal. However curare is known to have direct effects on such physiological parameters as excitatory receptive field separation between ipsilaterally and contralaterally activated tectal units [Raybourn, 1975], and receptive field sizes of units in the nucleus isthmi [Gruberg, personal communication]. In addition to these direct effects, the use of curare is very likely to inhibit lens accommodation. Such inhibition could, by itself, induce changes in the light pattern falling on the retina and elicit neural activity.

Besides using drugs to paralyze or quiet experimental animals, most electrophysiological experiments are done with the animals firmly mounted to an apparatus, so that no head or body motion is possible. It is well known that freely functioning frogs and toads, even when in a stationary pose, make eye movements that are synchronized with their breathing [Schipperheyn, 1963]. These mounting techniques effectively eliminate any motion of the eyes induced by body movements.

In short, experimental techniques are not adequate to allow the inference that freely functioning frogs and toads are essentially blind unless there is motion in the visual field. It is our opinion that in visual animals such as frogs and toads, the visuomotor system is highly integrated, so that disruptions in motor function will strongly influence visual function.

6

Conclusions

In our introductory chapter we expressed the hope that, by modeling a natural system, we could make contributions to the fields of both brain theory and robotics. Specifically, we wished to contribute to the understanding of the depth perception process in frogs and toads and, at the same time, to suggest depth algorithms that would be useful in the design of robotic systems. But our initial hope must be tempered by the fact that the disciplines under consideration are both in early and exploratory stages. Except under highly constrained conditions, robots do not yet make successful use of sensory feedback to control their actions. Similarly, attempts to relate animal behavior to underlying neurophysiology and neuroanatomy entail the making of assumptions that are often untestable using current methods. It is clear that, at this stage of development, a study such as ours can make only preliminary claims.

Some would argue, then, that work of this nature is premature, that successful modeling must await major advances, particularly in the neurosciences. Here we disagree strongly. The brain is so complex, and experimentation is so difficult, that progress in the neurosciences cannot be made without guidance from theory. Theory, on the other hand, is often far too abstract to provide the kind of concrete hypotheses that can be readily subjected to experimental testing. Modeling, particularly when it is accompanied by simulation, is a process of transforming abstract theory into testable structures. To the extent that the structure and performance of a model conform to current theory and available experimental data, they act as a confirmation of both the theory and the data. Where structural constraints from experimental data are lacking, the modeler must make assumptions that, by themselves, are a rich source of hypotheses for experimental testing. Further, simulation results provide a basis for predictions about the expected performance of the modeled system. This is especially important when performance data are lacking or difficult to obtain. In turn, as better theory is developed or new experimental data become available, a model must be updated, refined ,or replaced. We feel that the attainment of lasting and solid results depends upon a continuing cycle of theory, modeling, and experiment; biological investigation, and machine simulation. To slight any one of these avenues must seriously hamper activity in the other areas. The proper role for modeling, especially at this stage, is to provide enough of a synthesis of the information from the other disciplines to allow the formation of hypotheses that are capable of guiding further experimentation.

This final chapter reflects the role that we envision for modeling. It very briefly reviews our models but, more importantly, it outlines a series of animal experiments suggested by the simulations, and algorithms that bear consideration for application in robotic systems. We feel that these are the most important contributions of the work. The final section of this chapter outlines fruitful paths for future modeling efforts.

6.1 The Models and their Contributions

The most salient result of the modeling work is that we have demonstrated ways in which multiple depth cues can be used to enhance the speed and reliability of depth perception. Motivated by data from the literature on frogs and toads, we chose binocular disparity and lens accommodation as the two sources of depth cues for our study. To our knowledge, this is the first time that models of depth perception have been constructed using these two cues together.

The best way to use these two depth cues is to exploit the differences in their characteristics, rather than to simply take a weighted average of separate depth estimates based on each cue. Depth cues from binocular disparity are very precise, but suffer from inherent ambiguity; it is often difficult to determine which of several cues is the correct one. On the other hand, depth cues from accommodation are much less precise, and are subject to variation in accuracy with change in pupil diameter, but they are not ambiguous. Both of our models capitalize on the accuracy of binocular cues and the lack of ambiguity in the accommodation cues. Accommodation cues are used, in the models, to bias binocular cues so that their ambiguity is removed. They are not used (or are used with a low weight compared with binocular cues) to compute the final depth estimate. In this way, the full binocular accuracy is maintained.

Our models demonstrate two different ways in which accommodation biasing can be done. In the *cue interaction model* of Chapter 3, a passive process is envisioned. A record of image intensity as a function of lens setting is maintained at each retinal position. This record provides one of the inputs to the model. The other input is a map of matches between the binocular images over a range of disparities. These inputs are used to drive two separate processes that share information so that they build nearly identical depth maps of the binocular visual field. We say that the model is passive because it takes no special actions based on the visual input that it receives. Its structure is such that its depth-mapping activity is the same, no matter what is going on in front of the eyes. In contrast, the *prey localization model* of Chapter 4 is based on an active process that adjusts the state of lens accommodation in direct response to the pattern of visual input. In this model, prey selectors on each side of the visual system supply the accommodation controller with the retinal coordinates of a selected object. The controller then adjusts the lens based on the disparity between the object's positions on the right and left retinas. This mechanism acts to reject binocularly mismatched selections. A mismatch will cause a severe defocusing of the lenses. This defocusing will alter the input to the prey selectors, eventually correcting the mismatch. Correct binocular matches will improve lens focus, thus enhancing the inputs to the prey selectors, and confirming the match.

In our modeling work we came to the conclusion that depth perception, at least in frogs and toads, is probably not a single process. Our current point of view is that, since different visuomotor tasks require different depth information, there is no reason to assume *a priori* that there is only one depth perception system.

The two different models are each specialized for providing the depth information necessary to support a particular visuomotor activity. The *cue interaction model* is well suited to tasks involving navigation through the environment, and the *prey localization model* is especially appropriate for prey catching. The passive process of the *cue interaction model* is slower than the active process of the *prey localization model*, but it produces a complete depth map. Navigation appears to require a complete depth map, but is generally concerned only with stationary objects, so that this map can be built using a relatively slow process. Prey catching, on the other hand, requires the location of only a single point, but the location of that point must be obtained rapidly and with high reliability. The active process of the *prey localization model* is well suited to this task.

If lens accommodation is used in conjunction with binocularity for depth perception, and if depth perception processes are specialized to support different kinds of activity, then there must be a high degree of integration between the visual and motor systems. That vision should have a strong effect on motor activity comes as no surprise, but what has received very little attention is the effect that motor activity may have on vision. The extended model of accommodation control, described in Chapter 5, is an initial attempt at addressing some of the complexities of the interactions between the visual and motor systems.

6.2 Suggestions for Animal Experiments

The studies done with computer simulations of the models led to several interesting predictions, and also raised several important questions. The following discussion suggests experiments both to test the predictions and to resolve the questions.

Our simulation of the *cue interaction model* indicates that accurate mapping of the depth of barrier-like objects will be dependent upon a high degree of correspondence between monocular and binocular depth cues. This interdependency is much less marked when very sparse input from prey-like objects is used. However, the only experimental data showing the interaction of binocularity and accommodation come from prey-catching studies. The sensitivity to mismatch between cues, in determining barrier depth, suggests that placing either lenses or prisms in front of the eyes of a frog or toad will greatly disturb its depth estimates for navigation. Thus, a useful set of behavioral experiments would be to observe toads, with lenses or prisms mounted in front of their eyes, negotiating a fence-like barrier. These observations could be analyzed to look for differences in depth-mediated behavior between control animals, animals with lenses, and animals with prisms. It would be interesting to see if the complex perturbations of the estimates of the depth of barriers noted in our simulation are exhibited by frogs or toads. The simulation predicts that concave lenses will have a consistent effect, with stronger lenses causing fences to appear closer to the animal than do weaker lenses. Prisms, on the other hand, should have a more variable effect, even causing the fence to appear fragmented, so that the animal might occasionally behave as if there were a gap in the fence. To date, experiments of this sort have not been attempted due to the difficulty of motivating toads to negotiate barriers under the necessary experimental conditions [Collett, personal communication].

The simulation of the *cue interaction model* takes considerably longer to converge when binocular cues are removed, leaving only accommodation cues. Although we have no simulation results to show this, it can be inferred that the extended accommodation control system of Chapter 5 would also be slower with binocular input removed, since this controller depends on binocular input to achieve a rapid approximate lens setting. Thus, we hypothesize that monocular blinding of frogs and toads will increase their snapping latency. Experiments to test this hypothesis should be easy to perform.

The simulation of the *prey localization model* showed a marked increase in convergence time when either simulated lenses or prisms were used. This is because the total gain of the system is reduced when binocular and monocular cues are out of correspondence. Thus, a good test of the way in which we envision that binocular and accommodation cues interact for prey catching would be to conduct timed prey-catching experiments with frogs or toads to see if they exhibit an increase in snapping latency when lenses or prisms are used.

The structures of both the *cue interaction* and the *prey localization* models are such that they would be much more responsive to lateral movements than to movements away from or toward the animal. The *cue interaction model* computes the spatial position of an object by using retinal position to index a calculated depth-segmentation of the visual field. Since small lateral movements would have only an angular effect, recomputation of spatial location would be merely a matter of changing the retinal position used to index the segmentation. A similar result obtains for the *prey localization model*. This is because lateral movements would require only small adjustments in lens accommodative state, as opposed to the much larger adjustments that would be required to track an object moving toward or away from the eyes. Thus we predict that frogs and toads should be much more responsive and accurate in compensating for lateral movement than for other movements. An interesting comparison may be made between this prediction and the results of Westheimer and McKee [1978] who found that stereoscopic depth perception in humans is considerably more accurate when an object is moved across the visual field than when its movement is in depth.

The *prey localization model* is based on the use of accommodation to disambiguate binocular depth cues, but ambiguity is a problem only when more than one prey object is present in the visual field. It is not possible to cause this model to make mismatch errors when only one prey object is presented. Therefore, a definitive test of the way in which the *prey localization model* utilizes monocular depth cues would be to observe prey-catching behavior in the presence of multiple stimuli in animals whose accommodative mechanism has been disabled. Jordan et al. [1980] have shown that binocular toads whose lens accommodation muscles have been treated with either a muscle relaxant or a muscle contractor have little trouble estimating the depth of single targets presented in the frontal binocular field. However, we predict that if similarly treated animals are presented with multiple targets they will exhibit a marked tendency toward making binocular correspondence errors, and will often snap at "ghost" targets rather than at real targets.

To move beyond our preliminary models will require data from further animal experiments. There is still abundant literature on depth perception in frogs and toads with regard to the perception of the depth of prey. However, although there is ample evidence that these animals are able to judge accurately the depth of stationary barrier-like objects [Collett, 1982], very little is known about the mechanisms underlying this form of depth resolution. It is quite certain that frogs and toads determine the depth of prey using both binocularity and lens accommodation, but the depth cues used in barrier depth estimation are unknown. Another gap in the literature concerns the lens accommodation mechanism itself. There are, as yet, no data available showing the accommodation process in action. In Chapter 5, our model of accommodation control was built to address overt behavioral data, but a truly accurate description of this control will require data showing the accommodative response of the lenses when various visual configurations are presented.

Experiments with frogs or toads should be performed to test the hypothesis that both binocular and image focus cues are utilized to control lens accommodation. If a means were available to directly measure the accommodative state of the lenses, a test could be made to determine whether prisms have any effect on this state. If there is an effect from prisms, then it will be shown that binocular depth cues are being utilized. Also, if binocular cues are used to control accommodation, then lenses placed in front of the eyes of binocular frogs or toads will have a much smaller effect on accommodation than will lenses placed in front of the eyes of monocular animals.

To develop a more definitive model of lens accommodation than we have provided in Chapter 5, it will be necessary to obtain data to answer the following questions in both binocular and monocular animals:
(1) When there is no stimulus present in the visual field, does the lens scan in and out as if "searching" for a stimulus, or does it stay in a rest position?
(2) When there are several barrier surfaces in the visual field, does the lens intermittently focus on each of the surfaces in an exploratory way, or is the behavior similar to that when there is no stimulus?
(3) When there is a prey object and no other stimulus in the visual field, how does the accommodation process differ from that observed when there is no stimulus?
(4) When there are barrier surfaces in the visual field and a prey object is introduced, how does the accommodation process differ from that observed when there are only barriers?
To provide answers to these questions, an experimental methodology must be developed that will allow the observation of the movement of the lens or, alternatively, the tension in the accommodation muscles, as the animal is confronted with different visual configurations.

The model of accommodation control presented in Chapter 5 is too tentative to allow us to make firm predictions about the probable outcome of the experiments suggested above, but we can list findings that would be confirmatory to its structure. If our model is correct, then when confronted with a scene involving barriers but no prey stimuli, toads will scan their lenses in an oscillatory manner. Further, this oscillatory scanning will be momentarily interrupted as each barrier is brought into clear focus. If focus cues control the lenses separately for each eye, then the accommodative state at which the two lenses stop will differ if the scene is bilaterally asymmetric. When a prey-like stimulus is introduced into the binocular visual field, our model envisions that both lenses will rapidly accommodate to a state near to the depth of the prey, and then enter a slower fine-focusing phase.

6.3 Suggestions for Robotic Algorithms

Both depth perception models differ from previous theoretical models in that they involve adjusting the image sensors to obtain depth cues. For this reason, our models were not tested using static stereoscopic images, but instead were tested against an input scene that included the depth dimension. In this sense, both models are more suitable than models designed to process static images for representing the depth resolution process of animals. For this same reason, the models are also more appropriate for application in robotics. Robots interact not with static images but with the three-dimensional world, and thus could benefit from the greatly enhanced speed and reliability resulting from the use of multiple depth cues, some of which are obtained from sensor adjustment.

The method of using accommodation cues to disambiguate binocular cues that is outlined by the *cue interaction model* might be useful in robotic systems where the primary visual task is to maintain a map of the stationary objects in the visual environment. Such a scheme could employ a slow oscillatory scanning of the lenses of two imagers to provide image focus input, and disparity matching to provide binocular cues. Because of its organization, the algorithm underlying the *cue interaction model* would be very fast if implemented in an array processor.

The *prey localization model* has even more obvious applications in robotics. Given a limited class of objects to localize and a controlled visual environment, this model could be used as the basis for a simple but powerful object localization scheme. Its internal structure would allow easy implementation in parallel hardware for real-time performance. Additionally, its layers could be programmed as separate modules running independently in separate computing units. These modules could then be individually tuned to suit a wide variety of applications. For instance, the pattern recognizers could be tailored to recognize application-specific objects without modifying either the imagers, the selectors, or the accommodation controller.

A side benefit that would come from implementing the models in a robotic system would be to provide data that would help to further refine the models. The models have, as yet, been tested only on simple simulated two-dimensional scenes. Implementing them to interact with a three-dimensional environment and with direct video input is bound to uncover complications and possibilities that were not envisioned in our study.

Appendix A

Modeling and Simulation Details

The cooperative-competitive computational mechanism used in both the *cue interaction* and the *prey localization* models had its origin in an analytical paper by Amari and Arbib [1977]. Their analytical format is reflected in the fact that both models are represented by continuous equations and solved numerically, rather than by difference equations.

The analytical representation is based upon a definition of a single idealized neural unit. This definition is then extrapolated over a continuous homogeneous layer. Integro-differential equations are used to describe a layer's internal potential. In this appendix we explain the development of this general representation, and provide details of the computer simulations of both depth models.

A.1 Representation of a Neural Unit

Our representation of a single neural unit consists of (1) a set of weighted inputs, (2) an internal potential, (3) a differential equation that relates change in internal potential to both the current potential and the strength of the inputs, and (4) a function that converts internal potential to an external potential or firing rate.

We assume that the differential equation relating the inputs to the internal potential is both first order and linear, with time constant $\mathcal{T}$. Thus, the equation describing internal potential is

$$\mathcal{T}\dot{M} = -M + \sum_{i=1}^{n} K_i I_i \tag{A.1}$$

where M represents internal potential; n is the number of inputs; I_i, $1 \leq i \leq n$, is the ith input; and K_i is the ith input weight.

Two different functions are used to convert internal potential M to external firing rate. The first of these is a sigmoidal saturation-threshold function (Fig. A.1a) and is used in the representation of excitatory units. Inhibitory units employ a simple thresholded linear function (Fig. A.1b). The symbol f, with appropriate subscripts, is used throughout to represent a saturation-threshold function, and the symbol g to represent the thresholded linear function.

The general form of the saturation-threshold function used in all of the computer simulations was chosen for convenience of the interactive com-

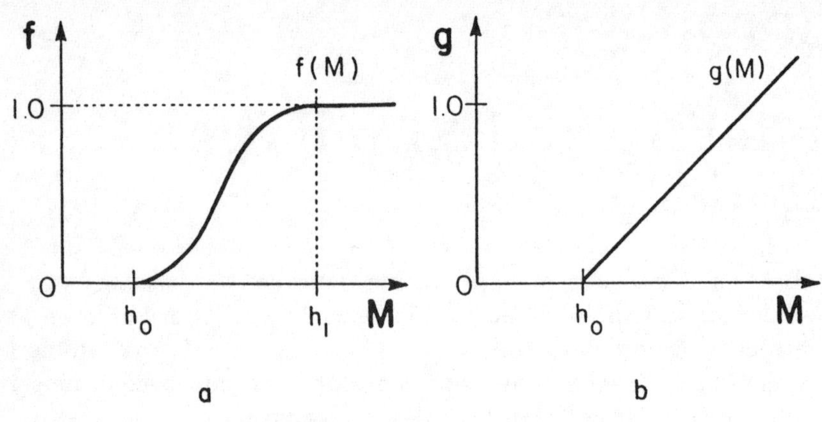

FIGURE A.1. Threshold Functions

puter graphics interface to the modeling software. It is simply a cubic spline function between the threshold and saturation levels, and was derived by constraining both the function and its first derivative to be continuous. The saturation threshold function is described by

$$f(M) = \begin{cases} 0, & M < h_0 \\ 3[(M-h_0)/\sigma]^2 - 2[(M-h_0)/\sigma]^3, & h_0 \le M \le h_1 \\ 1, & M > h_1, \end{cases} \tag{A.2}$$

where $\sigma = h_1 - h_0$, h_0 is the threshold, and h_1 is the saturation level.

A particular instance of the saturation-threshold function is completely defined by specifying h_0 and h_1.

The thresholded linear function is completely defined by specifying its single threshold, h_0. Its formulation is given by

$$g(M) = \begin{cases} 0, & M < h_0 \\ M - h_0, & M \ge h_0. \end{cases} \tag{A.3}$$

A.2 Representation of a Neural Layer

A neural layer is a continuous approximation to an array of individual neurons, constructed by first replicating the definition of a neural unit over a continuous one- or two-dimensional surface, and then defining the connectivity between loci on the layer. A layer is homogeneous in all of its properties. For a one-dimensional layer we let $w'(x,y)$ represent the weight of the influence of a cell at position y on a cell at position x. We constrain this weighting function by requiring that the strength of the connections between two points be a function only of the Euclidian distance between points x and y, i.e. $|x-y|$. Therefore, this connectivity may be represented by an even function w, defined by

$$w(x-y)=w'(x,y),$$

with

$$w(s)=w(-s).$$

For a two-dimensional layer, connectivity between two points is constrained to be a function of the magnitudes of the two components r and s of the vector difference between the two points. Thus, connectivity is not radially symmetric but does have the property

$$w(r,s)=w(\pm r,\pm s).$$

If we let $M(x,y,t)$ represent the internal point potential at position (x,y) and time t in two-dimensional layer M, and let $f[M(x,y,t)]$ be the firing rate at that point, then the strength of the stimulation received by a point from its neighboring points is given by the convolution

$$\int\int w(x-\zeta,y-\eta)f[M(\zeta,\eta,t)]d\zeta d\eta.$$

The complete description of the point potential in a two-dimensional layer M is given by

$$\mathcal{T}\dot{M}(x,y,t)=-M(x,y,t)+\int\int w(x-\zeta,y-\eta)f[M(\zeta,\eta,t)]d\zeta d\eta + \sum_{i=1}^{n}K_iI_i(x,y,t), \quad \text{(A.4)}$$

where $I_i(x,y,t)$, $1\leq i\leq n$, represents external input. The external firing rate at a point on a two-dimensional layer is given by either

$$f[M(x,y,t)] \text{ or } g[M(x,y,t)],$$

with external firing rate $f[U(x,t)]$ or $g[U(x,t)]$.

Similarly, a one-dimensional layer U is described by

$$\mathcal{T}\dot{U}(x,t)=-U(x,t)+\int w(x-\zeta)f[U(\zeta,t)]d\zeta+\sum_{i=1}^{n}K_iI_i(x,t). \quad \text{(A.5)}$$

The functions w that govern the lateral spread of excitation within a layer are represented in the simulations by piecewise polynomial functions, again for convenience of the interactive graphics interface. All of the spread functions used were of the form shown in Fig. A.2. Neither of the models required the simulation of a two-dimensional spread function. In Fig. A.2 the s axis represents the difference between the positions of two points. The points $(0,w_0)$, (s_1,w_1), and $(s_2,0)$ represent three knot points through

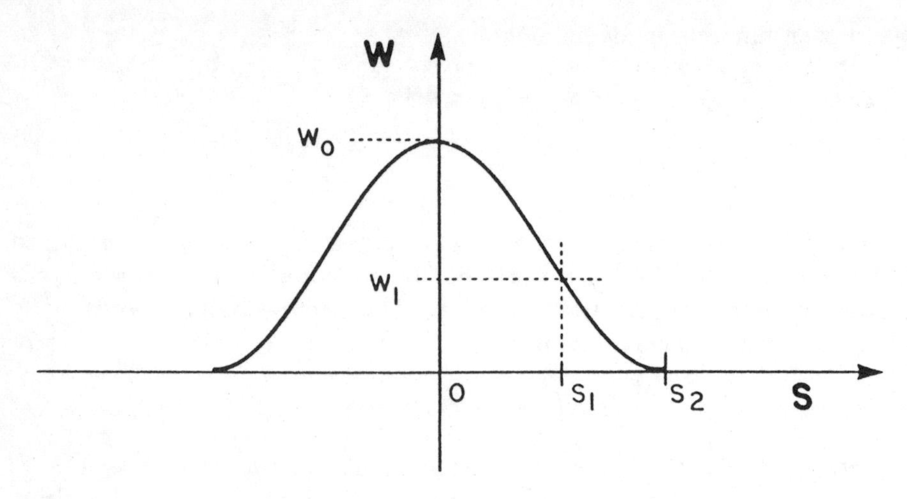

FIGURE A.2. Typical Spread Curve

which the piecewise polynomial is drawn. We represent the portion of the curve between 0 and s_1 by a quadratic and the portion between s_1 and s_2 by a cubic. We constrain the curve by requiring it and its first derivative to be everywhere continuous. The description of w is then

$$w(s) = \begin{cases} (w_1 - w_0)(|s|/s_1)^2 + w_0, & 0 \leq |s| \leq s_1 \\ C_1[(s_2 - |s|)/\sigma]^3 + C_2[(s_2 - |s|)/\sigma]^2, & s_1 < |s| < s_2 \\ 0, & |s| \geq s_2, \end{cases} \tag{A.6}$$

where $\sigma = s_2 - s_1$, $C_1 = w_1 - C_2$, and $C_2 = 3w_1 + 2(w_1 - w_0)(s_2 - s_1)/s_1$. Thus, a particular instance of the spread function w is completely specified by the four parameters w_0, w_1, s_1, and s_2.

A.3 Numerical Methods

Both models are qualitative representations of computational processes and do not attempt to replicate the detailed functioning of known physical systems. Further, for the most part we were not interested in the transient performance of the simulations, but only in their equilibria. Therefore, we did not use sophisticated numerical techniques for the simulations. Instead, we chose to use simple approximation methods. The spatial integrals in the equations describing the models were computed using the trapezoidal rule, and the differential equations were solved using Euler's method. To insure that the use of Euler's method did not seriously degrade accuracy we chose the time increments for the simulations carefully. This was done by running the simulations against standard test scenes using various time increments, and noting the point at which reduction of the time increment ceased to have a noticeable effect on the resulting equilibrium state. All simulations were begun from an initially inert state (zero potential in all layers), and excitation was taken to be zero beyond the boundaries of a layer when computing the spatial convolution integrals.

A.4 The Cue Interaction Model

In the computer simulation of the *cue interaction model* of Chapter 3 the monocular layer M and the stereoptic layer S were modeled as 41×41 arrays. The corresponding inhibitory layers U and V were modeled as vectors of length 41. The binocular disparity input plane D and the accommodation input plane A were also modeled as 41×41 arrays. These inputs were calculated as described in Appendix B and, when required, included the effect of simulated interposed lenses and prisms.

A reasonable starting set of parameters for the simulation was determined by an equilibrium analysis. The time derivatives in eqs. (3.3) were set to zero to obtain the equilibrium equations

$$
\begin{aligned}
M(q,d) &= \int w_m(q-\zeta)f[M(\zeta,d)]d\zeta + K_{sm}f[S(q,d)] \\
&\quad -K_m g[(U(q)] + K_a A(q,d), \\
U(q) &= K_u \int f[M(q,\eta)]d\eta, \\
S(q,d) &= \int w_s(q-\zeta)f[S(\zeta,d)]d\zeta + K_{ms}f[M(q,d)] \\
&\quad -K_s g[(V(q)] + K_d D(q,d), \\
V(q) &= K_v \int f[S(q,\eta)]d\eta.
\end{aligned}
\tag{A.7}
$$

These equilibrium equations were simplified by applying several *ad hoc* approximations and constraints. First, the equations were approximated by discrete equations, with indices i and j replacing spatial coordinates q and d, summations replacing integrals, and discrete steps Δq and Δd replacing the differentials. With the exception of the input gains K_a and K_d, corresponding gains and spread functions were constrained to be identical across the two pairs of equations. Cross-coupling gains K_{sm} and K_{ms} were replaced by K_c, inhibitory feedback gains K_m and K_s were replaced by K_i, and inhibitory layer input gains K_u and K_v were replaced by K_o. Similarly, spread functions w_m and w_s were replaced by the single function w. The extent of the summations used to approximate the spatial convolutions were constrained to three discrete spatial steps. This allowed the spread function w to be reduced to two constants W_0 and W_1, with W_0, the integral of w from $-\Delta d/2$ to $\Delta d/2$, representing the central portion of w;

TABLE A.1. Equilibrium Solution Constraints

$$
\begin{array}{llcllcll}
(1) & M(i,j) & > & h & \rightarrow & M(i,k) & < & h, \quad k \neq j \\
 & S(i,j) & > & h & \rightarrow & S(i,k) & < & h, \quad k \neq j \\
(2) & M(i,j) & > & h & \rightarrow & M(i,j) & \geq & h+1 \\
 & S(i,j) & > & h & \rightarrow & S(i,j) & \geq & h+1 \\
(3) & M(i,j) & > & h & \rightarrow & S(i,j) & > & h \\
 & S(i,j) & > & h & \rightarrow & M(i,j) & > & h
\end{array}
$$

and W_1, the integral of w from $\Delta d/2$ to ∞, representing the two portions of w to either side of the central portion. The application of all of these simplifications to eqs. (A.7) gives

$$
\begin{aligned}
M(i,j) &= W_0 f[M(i,j)] + W_1\{f[M(i-1,j)] + f[M(i+1,j)]\} \\
&\quad + K_c f[S(i,j)] - K_i g[(U(i)] + K_a A(i,j), \\
U(i) &= K_o \Delta d \sum_k f[M(i,k)], \\
S(i,j) &= W_0 f[S(i,j)] + W_1\{f[S(i-1,j)] + f[S(i+1,j)]\} \\
&\quad + K_c f[M(i,j)] - K_i g[(V(i)] + K_d D(i,j), \\
V(i) &= K_o \Delta d \sum_k f[S(i,k)].
\end{aligned}
\tag{A.8}
$$

The analysis was further simplified by *a priori* selection of the saturation level of function f, and the threshold of function g. We let h represent the threshold of f and defined its saturation level to be $h+1$. The threshold of function g was defined to be zero.

Assumptions were also made to constrain the form of the equilibrium solutions. First, we invoked the uniqueness constraint that, for each discrete retinal position i, at most one element of array M and one element of array S will have a level of excitation above threshold h. Further, we sought to enforce the constraint that, if an element $M(i,j)$ or $S(i,j)$ were above threshold, it would be in saturation. We also sought to require that if a point in one excitatory layer, for example, $M(i,j)$, were above threshold, then the corresponding point in the other layer, $S(i,j)$, would also be above threshold. These constraints are summarized in Table A.1. It follows from these constraints that, for a given retinal position index i, either no points of arrays M and S are above threshold h, or exactly one point is in saturation in each array and that these points have the same disparity index j.

It also follows from the constraints of Table A.1 and the definition of saturation function f that we may define a parameter

$$\Omega_{i,j} = f[M(i-1,j)] + f[M(i+1,j)] = f[S(i-1,j)] + f[S(i+1,j)],$$

that will take on values 0, 1, or 2 depending upon how many left and right neighbors of point (i,j) are above threshold. To take advantage of this, the continuity constraint of Chapter 3 was invoked in a very strict sense: if a point is in saturation then its neighbors are in saturation, and if it is not above threshold then its neighbors will not be above threshold. Thus we have

$$\Omega_{i,j} = \begin{cases} 2, & M(i,j) > h+1, S(i,j) > h+1 \\ 0, & M(i,j) < h, S(i,j) < h. \end{cases} \tag{A.9}$$

Because of the three constraints listed in Table A.1 and the choices made for parameters of threshold functions f and g, the equilibrium equations may take two possible forms. If there is a disparity index j at retinal-position i for which M and S are above threshold, then

$$\begin{aligned} M(i,k) &= \begin{cases} W_0 + 2W_1 + K_c - K_i K_o \Delta d + K_a A(i,k), & k = j \\ -K_i K_o \Delta d + K_a A(i,k), & k \neq j \end{cases} \\ U(i) &= K_o \Delta d, \\ S(i,k) &= \begin{cases} W_0 + 2W_1 + K_c - K_i K_o \Delta d + K_d D(i,k), & k = j \\ -K_i K_o \Delta d + K_d D(i,k), & k \neq j \end{cases} \\ V(i) &= K_o \Delta d. \end{aligned} \tag{A.10}$$

If, however, no point at retinal-position index i is above threshold, then

$$\begin{aligned} M(i,j) &= K_a A(i,j), \\ U(i) &= 0, \\ S(i,j) &= K_d D(i,j), \\ V(i) &= 0. \end{aligned} \tag{A.11}$$

Eqs. (A.10), together with the various constraints, serve to place bounds on the allowable parameter settings. In order for both the uniqueness constraint (1) and the saturation constraint (2) to hold, we must have

$$\begin{aligned} W_0 + 2W_1 + K_c - K_i K_o \Delta d + K_a A(i,j) &> h+1, \\ W_0 + 2W_1 + K_c - 2K_i K_o \Delta d + K_a A(i,j) &< h. \end{aligned}$$

These inequalities may be subtracted to yield the restriction on net inhibitory gain

$$K_i K_o \Delta d > 1. \tag{A.12}$$

In order for the similarity constraint (3) and the saturation constraint (2) to hold, we must have

$$\begin{aligned} W_0 + 2W_1 + K_c - K_i K_o \Delta d + K_a A(i,j) &> h + 1, \\ W_0 + 2W_1 - K_i K_o \Delta d + K_a A(i,j) &< h. \end{aligned}$$

These inequalities may be subtracted to yield the restriction on the cross-coupling gain

$$K_c > 1. \tag{A.13}$$

In order for both the continuity constraint and the saturation constraint (2) to hold, we must have

$$\begin{aligned} W_0 + 2W_1 + K_c - K_i K_o \Delta d + K_a A(i,j) &> h + 1, \\ W_0 + K_c - 2K_i K_o \Delta d + K_a A(i,j) &< h. \end{aligned}$$

These inequalities may be subtracted to yield the restriction on the non-central portion of the excitatory spread function

$$W_1 > 0.5. \tag{A.14}$$

By adding an additional constraint that the model be able to work in a solely monocular mode, we obtain the other needed restriction on the excitatory spread function. Setting K_d to zero in eq. (A.10) and invoking the similarity constraint (3) and the saturation constraint (2) gives

$$W_0 > h + 1 + K_i K_o \Delta d - K_c - 2W_1. \tag{A.15}$$

The choice of input gains K_a and K_d were restricted by applying a further constraint that there be threshold input values A_0 and D_0 such that

$$\forall i [\forall j M(i,j) < h] \rightarrow [\forall j A(i,j) < A_0],$$

and

$$\forall i [\forall j S(i,j) < h] \rightarrow [\forall j D(i,j) < D_0].$$

Eqs. (A.11) hold only when all elements of M and S at retinal position i are below threshold h, so together with the input threshold constraint they require

$$\begin{aligned} K_a &< h/A_0, \\ K_d &< h/D_0. \end{aligned} \tag{A.16}$$

It is clear that not all of the constraints applied in the derivation of inequalities (A.12)-(A.16) can hold everywhere. However, they served to simplify the analysis enough to allow us to make an educated guess at a reasonable set of initial parameter settings. The most important constraint is uniqueness (1), so we chose the inhibitory gain to be 80% greater than the lower limit given by inequality (A.12). The maximum allowable disparity represented in the simulation was ± 0.75, and the disparity coordinate was represented by 41 discrete steps (indices -20 to 20) so that the disparity step size Δd was 0.0375. The inhibitory layer input gain K_o was set to 80.0, and the inhibitory feedback gain K_i was set to 0.6. This gave a net inhibitory gain $K_i K_o \Delta d$ of 1.8. The similarity and continuity constraints were less important. They were weakly enforced by setting the cross-coupling gain and the non-central portion of the excitatory spread function to the borderline values of inequalities (A.13) and (A.14): K_c was set to 1.0, and W_1 was set to 0.5. The threshold h was arbitrarily chosen to be 0.1. After making these choices, the resulting borderline value of inequality (A.15) was used to set the central portion W_0 of the excitatory spread function to 0.9. Finally, the simulation was made more sensitive to binocular input than to monocular input by setting D_0, the threshold on disparity input, to 0.2, and A_0, the threshold on accommodation input, to 0.5. The borderline values of inequalities (A.16), were used to set the disparity input gain K_d to 0.5 and the accommodation input gain K_a to 0.2.

The equilibrium analysis of the *cue interaction model* gave us a starting set of simulation parameters. The nominal parameter settings used in the experiments were determined by making adjustments to this starting set both to improve transient performance and to better approximate behavioral data. These nominal parameter settings are shown in Table A.2. The only parameters requiring adjustment were the cross-coupling gain K_c, which was reduced 20% to 0.8, and the excitatory spread function, whose central portion W_0 was reduced 25% to 0.68 and whose non-central portion was reduced 50% to 0.25.

A.5 The Prey Localization Model

Most of the layers of the *prey localization model* of Chapter 4 are represented as one-dimensional homogeneous layers of the type described by eq. (A.5). Exceptions to this are the inhibitory cell layers, which are modeled as single neural units as described in eq. (A.1); and the imager layers, which are simply the retinal inputs R_L and R_R described by eq. (B.3) in Appendix B. This model also contains an accommodation controller that is modeled as a lumped parameter system whose input is two sets of vector quantities.

All of the one-dimensional layers of this model were represented in the computer simulation as vectors of length 41. The accommodative input planes A_L and A_R were represented by 41×41 arrays. These inputs were calculated from the retinal inputs R_L and R_R as described in Appendix B

TABLE A.2. Cue Interaction Model Parameter Settings

Simulation Time Increment, $\Delta t = 0.05$

Time Constants

$\mathcal{T}_m = 0.30 \quad \mathcal{T}_u = 0.10 \quad \mathcal{T}_s = 0.30 \quad \mathcal{T}_v = 0.10$

Saturation Functions

	h_0	h_1
f	0.10	1.10
g	0.00	- - -

Spread Functions

	w_0	w_1	s_1	s_2	W_0	W_1
w_m	17.78	7.11	0.0375	0.0750	0.68	0.25
			(3.375°)	(6.750°)		
w_s	17.78	7.11	0.0375	0.0750	0.68	0.25
			(3.375°)	(6.750°)		

Gains

$K_{sm} = 0.80 \quad K_m = 0.60 \quad K_u = 80.0 \quad K_a = 0.20$

$K_{ms} = 0.80 \quad K_s = 0.60 \quad K_v = 80.0 \quad K_d = 0.50$

and, when required, included the effect of simulated interposed lenses and prisms.

As with the *cue interaction model*, an equilibrium analysis was conducted in order to establish a reasonable set of parameters for the simulation. This analysis was begun by setting the derivatives in eqs. (4.1) - (4.7) to zero to obtain equilibrium equations, and then making a discrete approximation to the continuous equations. In the resulting discrete equations, all layers are represented by arrays with index i replacing the retinal-position coordinate q, and index j replacing the disparity coordinate d. The equations were further simplified by making several approximations. Each spread function w was replaced by a constant W that represents the integral under the function w for one retinal-position increment Δq about its center. To justify this step, it was assumed that both w_t and w_b are narrow with respect Δq, and that w_i, although it is broad with respect to Δq, is nearly constant over its width. The thresholds h_0 for the saturation-threshold functions f_t and f_b were arbitrarily set to 0, and the saturation levels h_1 were set to 1. The threshold for the threshold function g was set to 0. Also, the subscripts L and R, indicating the right and left sides of the visual system, were dropped since the analysis required only one set of equations.

Given these simplifications, the equilibrium solution to eq. (4.1), governing the pattern recognizers, is

$$T(i) = W_t f_t[T(i)] + K_{at} A(i,j).$$

We wanted the level of excitation T in the pattern-recognizers to exactly reach saturation with the accommodation input at its maximum value of 1.0. Thus, K_a and W_t are related by

$$K_{at} = 1 - W_t. \tag{A.17}$$

We chose a value of 0.75 for W_t and 0.25 for K_{at}.

The equilibrium solution to eq. (4.2), governing the binocular layer, is

$$B(i) = W_b f_b[B(i)] + 2I(i) + K_{tb} f_t[T(i)] - K_{ub} U.$$

To assure robust performance, we wanted the binocular layer to reach a value 80% above saturation level when receiving full input from both the pattern-recognizer T and the two relay layers I. We also wanted to assure that in equilibrium the binocular cross-coupling through the relay layers would dominate the direct input from the pattern-recognizer. Thus, we set K_{tb} to 1/3 of the total relay input giving

$$K_{tb} = 0.67 I_{\max}, \tag{A.18}$$

where $I_{\max}$ is the maximum expected equilibrium level of the relay input. Applying these constraints and eq. (A.18) to the equilibrium equation yields

$$W_b = 1.8 + K_{ub} U - 2.67 I_{\max}. \tag{A.19}$$

The gain K_{ub} was arbitrarily set to 1.0, since the net inhibitory effect on the binocular layer could be controlled by parameters affecting the equilibrium level of the inhibitory cell U.

Eq. (4.3), the equation for the relay layers, reduces to

$$I = nW_i, \tag{A.20}$$

where n is the number of discrete points of the binocular layer that remain excited in the equilibrium. Experience with the simulation system showed that an appropriate maximum value for n was 2. W_i was arbitrarily given the value 0.2. This value was chosen in order to keep the excitation level in the relay layer in the range from 0 to 1. With these values for W_i and n, the maximum expected level of relay excitation I_{max} is 0.4.

The equilibrium equation for the inhibitory cell U is

$$U = nK_{bu}\Delta q. \tag{A.21}$$

In the simulation the retinal-position increment Δq was 0.0375. The value 20 was chosen for K_{bu} giving U the value 1.5 when 2 points remain excited in the binocular layer.

Applying the various parameter choices to eqs. (A.18) and (A.19) gives K_{tb} the value 0.27, and W_b the value 2.23.

Parameters were experimentally adjusted from these starting values in order to improve transient performance. The resulting nominal parameter settings are given in Table A.3.

TABLE A.3. Prey Localization Model Parameter Settings

Simulation Time Increment, $\Delta t = 0.025$

Time Constants

$\mathcal{T}_a = 0.40 \quad \mathcal{T}_t = 0.05 \quad \mathcal{T}_b = 0.10 \quad \mathcal{T}_u = 0.05$

Saturation Functions

	h_0	h_1
f_t	0.00	1.00
f_b	0.05	1.05
g	0.00	- - -

Spread Functions

	w_0	w_1	s_1	s_2	W
w_t	27.0	12.0	0.015 (1.35°)	0.030 (2.70°)	0.75
w_b	89.0	38.5	0.013 (1.17°)	0.030 (2.70°)	2.50
w_i	4.8	4.8	0.080 (7.20°)	0.125 (11.25°)	1.80

Gains

$K_{tb} = 0.25 \quad K_{at} = 0.25 \quad K_{bu} = 20.0 \quad K_{ub} = 1.00$

Appendix B

Simulation Optics

In this appendix we describe the equations of the optical system used to provide input to both models. We describe the translation from spatial to retinal coordinates, outline the method used to simulate lenses and prisms in front of the eyes, and define the method used to calculate initial depth estimates from disparity and accommodation.

B.1 Optical Geometry

The optical equations are all based upon the configuration depicted in Fig. B.1. Cartesian measurements are made relative to the x and y coordinates indicated by the bold orthogonal lines crossing in the center of the figure. The pupils of the two eyes are indicated by the bold dots near the bottom of the figure. The optical axis of each eye is indicated by the solid line passing through the pupil and labeled OA. The two optical axes intersect at the point $(0,y_f)$ indicated by the star. The retinas are indicated by the semicircles behind the pupils. The diagonals passing through the pupils

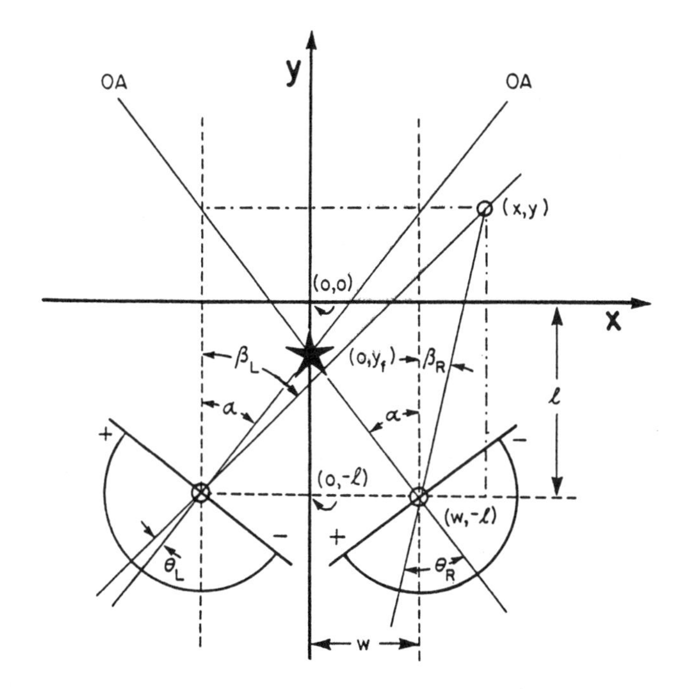

FIGURE B.1. The Optical System

are drawn orthogonal to the optical axes. Retinal angles, θ_L for the left eye and θ_R for the right eye, are measured relative to the optical axis, with sign as indicated at the ends of the retinal semicircles. The point (x, y) is an arbitrary spatial reference point, and the dashed lines are reference lines used in the derivations. The symbol l denotes the distance of the interpupillary line from the origin of the coordinate system, and w represents one-half of the interpupillary distance. α is the angle of declension of the optical axis from the ordinate. Similarly, β_L and β_R are the angles of declension of the rays passing through point (x, y) and, respectively, the left and right pupils.

Although the eyes of frogs and toads are outward facing (their optical axes cross behind the eyes) we employed an optical model with eyes facing towards a frontally located fixation point. This was done to simplify the simulation system, and is consistent with the fact that binocular tectal projections have been shown to be in register with respect to a frontal horopter [Gaillard and Galand, 1980]. Further, the use of an unmoving fixation point has no effect upon the optical equations except to shift the center of the resulting retinal position versus disparity coordinate system; the fixation point in space maps to the origin of the internal coordinate system.

The retinal angles θ_L and θ_R of the projections of point (x, y) are given by

$$\theta_L = \beta_L - \alpha,$$

and

$$\theta_R = \beta_R + \alpha.$$

Inspection of Fig. B.1 shows that

$$\alpha = \arctan \frac{w}{y_f + l},$$

$$\beta_L = \arctan \frac{x + w}{y + l},$$

and

$$\beta_R = \arctan \frac{x - w}{y + l}.$$

Finally, retinal angle is converted to retinal position by simply multiplying by a magnification factor. We chose retinal position 1.0 to represent 90°. Thus, if a is the magnification factor, q_L is position on the left retina, and

q_R is position on the right retina, we have

$$\begin{aligned} a &= 2/\pi, \\ \alpha &= \arctan \frac{w}{y_f + l}, \\ q_L &= a(\arctan \frac{x+w}{y+l} - \alpha), \\ q_R &= a(\arctan \frac{x-w}{y+l} + \alpha). \end{aligned} \tag{B.1}$$

B.2 Representation of Prisms

The presence of prisms in front of the eyes was simulated by applying an offset, proportional to prismatic strength, to the retinal coordinates determined by eqs. (B.1). We were interested in the effect of prisms in altering the disparity (positional difference) between the projections of a point on the left and right retinas. Thus, we expressed the prismatic strength parameter K_p as a percentage of the maximum disparity range $d_{\max}$ which we wished to consider. The prismatic offset C_p and the augmented equations for retinal position are given by

$$\begin{aligned} C_p &= \frac{1}{2} d_{\max}(K_p/100), \\ q_L &= a(\arctan \frac{x+w}{y+l} - \alpha) - C_p, \\ q_R &= a(\arctan \frac{x-w}{y+l} + \alpha) + C_p. \end{aligned} \tag{B.2}$$

B.3 Retinal Projections

The projection of visual data onto the retinas was simplified by the fact that the simulation software used to experiment with the models records the dimensions and location of each object in the visual space. The arrays representing the retinas were loaded by using eqs. (B.2) to determine the loci of the projections of each of these known objects onto the two retinas. Retinal points that received the projections of one or more objects were assigned the value 1. Points that received no projection were assigned the value 0. No attempt was made to account for occlusion of one object by another. Thus, if q is the retinal position coordinate, and R_L and R_R are the retinas

$$R_L(q) = \begin{cases} 0, & \text{no object projects to point q on left retina} \\ 1, & \text{any object projects to point q on left retina,} \end{cases}$$
$$R_R(q) = \begin{cases} 0, & \text{no object projects to point q on right retina} \\ 1, & \text{any object projects to point q on right retina,} \end{cases} \tag{B.3}$$

In the computer simulations the retinas were modeled as vectors of length 161.

B.4 Disparity Input Planes

Depth estimates based upon disparity were calculated from the retinal images and stored in planes organized according to a coordinate system based on retinal position q and disparity d. The binocular disparity inputs, D_L for the left side and D_R for the right side, were computed by the equations

$$D_L(q,d) = R_L(q) \cdot R_R(q+d),$$
$$D_R(q,d) = R_R(q) \cdot R_L(q-d), \tag{B.4}$$

The range of allowable disparities d_{max} was constrained to represent a shift of one retina over the other of no more than 25% of the binocular field in either direction. In the computer simulation the disparity planes were modeled as 41×41 arrays.

B.5 Accommodation Input Planes

The accommodation input planes, A_L and A_R, were also represented in the retinal angle q versus disparity d coordinate system. Depth estimates based upon lens accommodation were approximated by building each accommodation input plane via a rule which for each stimulated point on the corresponding retina produced a set of values distributed along the disparity coordinate. A Gaussian distribution centered about the correct depth value was assumed. Thus, the value at a point (q, d) in one of the accommodation input planes could be considered to be either an estimate of depth or a measure of the quality of focus of the image at retinal position q with the lens accommodated to a depth corresponding with disparity d. If point a, in space, projects to points q_L on the left retina and q_R on the right retina then the disparity of point a is defined to be

$$d_a = q_R - q_L.$$

The full range of accommodative depth estimates for the point a is given by

$$
\begin{aligned}
A_L(q_L, d) &= \exp[-(100\frac{d - d_a}{\sigma d_{\max}})^2/2], \\
A_R(q_R, d) &= \exp[-(100\frac{d - d_a}{\sigma d_{\max}})^2/2],
\end{aligned}
\tag{B.5}
$$

where σ is the spread parameter measured in percent of maximum disparity $d_{\max}$.

B.6 Representation of Lenses

A simulation of the presence of lenses in front of the eyes was provided by the introduction of a parameter K_l governing systematic error in the centering of the depth estimate curve about the true depth of a point. If this parameter is expressed in percent of maximum disparity then the adjusted center of the accommodative depth estimate curve is given by

$$d_a = q_R - q_L + d_{\max} K_l / 100. \tag{B.6}$$

Eqs. (B.5) and (B.6) together describe the method used to build the accommodation input planes. These planes were represented in the simulation as 41×41 arrays.

B.7 Conversion from Internal to External Coordinates

It was necessary in the simulations to be able to convert depth estimates expressed in the internal retinal angle versus disparity coordinate system back into the spatial coordinate system. From eqs. (B.2) and inspection of Fig. B.1 it can be demonstrated that, given a disparity d and a retinal position q_L on the left retina or q_R on the right retina, a unique spatial point (x, y) is determined. Once a pair of retinal angles is determined from either of the pairs of formulae

$$
\begin{aligned}
\theta_R = q_R/a, &\qquad \theta_L = (q_R - d)/a, \\
\theta_L = q_L/a, &\qquad \theta_R = (q_R + d)/a,
\end{aligned}
\tag{B.7}
$$

then the spatial coordinates are given by

TABLE B.1. Optical Simulation Parameter Settings

Parameter
y Coordinate of Fixation Point, $y_f = -10$ cm
Distance Between Pupils, $2w = 6.0$ cm ($w = 3.0$ cm)
Distance of Interpupillary Line from Origin, $l = 22$ cm
Retinal Angle to Retinal Position Conversion, $a = 2/\pi$
Accommodation Spread, $\sigma = \begin{cases} 25\% & \text{(cue interaction model)} \\ 50\% & \text{(prey localization model)} \end{cases}$
Prism and Lens Strength, $K_p = K_l = 0$
Maximum Disparity, $d_{\max} = 0.25$

$$\begin{aligned} y &= 2w/[tan(\theta_L + \alpha) - tan(\theta_R - \alpha)] - l, \\ x &= (y + l)tan(\theta_R - \alpha) + w. \end{aligned} \tag{B.8}$$

B.8 Nominal Parameter Settings

The nominal optical parameter settings used in all of the reported experiments are given in Table B.1.

Bibliography

Amari, S. and M. Arbib, Competition and cooperation in neural nets, in *Systems Neuroscience,* edited by J. Metzler, pp. 119–165, Academic Press, New York, 1977.

an der Heiden, U. and G. Roth, A mathematical network model for retinotectal prey recognition in amphibians, in *Proceedings of the Second Workshop on Visuomotor Coordination in Frog and Toad: Models and Experiments,* 1983, Technical Report 83-19, Computer and Information Science Dept., Univ. of Massachusetts, Amherst.

Arbib, M., Perceptual structures and distributed motor control, in *Handbook of Physiology — The Nervous System II. Motor Control,* edited by V. Brooks, pp. 1449–1480, Amer. Physiol. Soc., Bethesda, 1981.

Arbib, M. and D. House, Depth and detours: towards neural models, in *Proceedings of the Second Workshop on Visuomotor Coordination in Frog and Toad: Models and Experiments,* 1983, Technical Report 83-19, Computer and Information Science Dept., Univ. of Massachusetts, Amherst.

Autrum, H., Das fehlen Unwillkürlicher augenbewegungen beim frosch, *Naturwiss., 46,* 435, 1959.

Brown, W. T. and W. B. Marks, Unit responses in the frog's caudal thalamus, *Brain, Behav. Evol., 14,* 274–297, 1977.

Cervantes, F., R. Lara, and M. A. Arbib, A neural model subserving prey-predator discrimination and size preference in anuran amphibia, in *Proceedings of the Second Workshop on Visuomotor Coordination in Frog and Toad: Models and Experiments,* 1983, Technical Report 83-19, Computer and Information Science Dept., Univ. of Massachusetts, Amherst.

Collett, T., Stereopsis in toads, *Nature, 267,* 349–351, 1977.

Collett, T., Do toads plan routes? a study of the detour behavior of *Bufo viridis*, *J. Comp. Physiol., 146,* 261–271, 1982.

Collett, T. and L. I. K. Harkness, Depth vision in animals, in *The Analysis of Visual Behavior,* edited by D. Ingle, M. Goodale, and R. Mansfield, pp. 111–176, MIT Press, Cambridge, 1982.

Collett, T. and S. B. Udin, The role of the toad's nucleus isthmi in prey-catching behavior, in *Proceedings of the Second Workshop on Visuomotor Coordination in Frog and Toad: Models and Experiments,* 1983, Technical Report 83-19, Computer and Information Science Dept., Univ. of Massachusetts, Amherst.

Corvaja, N. and P. d'Ascanio, Spinal projections from the mesencephalon in the toad, *Brain, Behav. Evol., 19,* 205–213, 1981.

Dev, P., Perception of depth surfaces in random-dot stereograms: a neural model, *Int. J. Man-Machine Studies, 7,* 511–528, 1975.

Didday, R. L., *The Simulation and Modelling of Distributed Information Processing in the Frog Visual System,* PhD thesis, Stanford Univ., 1970.

Didday, R., A model of visuomotor mechanisms in the frog optic tectum, *Math. Biosci, 30,* 169–180, 1976.

Ewert, J., Single unit response of the toad (*Bufo americanus*) caudal thalamus to visual objects, *Z. Vergl. Physiol., 74,* 81–102, 1971.

Ewert, J., The visual system of the toad: behavioral and physiological studies on a pattern recognition system, in *The Amphibian Visual System: A Multidisciplinary Approach,* edited by K. Fite, pp. 141–202, Academic Press, New York, 1976.

Ewert, J., *Neuroethology,* Springer-Verlag, Berlin, 1980.

Ewert, J., Neuronal basis of configurational prey selection in the common toad, in *The Analysis of Visual Behavior,* edited by D. Ingle, M. Goodale, and R. Mansfield, pp. 7–45, MIT Press, Cambridge, 1982.

Ewert, J. and von Seelen, Neurobiologie und system-theorie eines visuellen muster-erkennungsmechanismus bei krote, *Kybernetik, 14,* 167–183, 1974.

Finch, D. J. and T. S. Collett, Small-field, binocular neurons in the superficial layers of the frog optic tectum, *Proc. R. Soc. Lond. B, 217,* 491–497, 1983.

Fischer, B., Overlap of receptive field centers and representation of the visual field in cat's optic tract, *Vis. Res., 13,* 2113–2120, 1973.

Fite, K. V., Single-unit analysis of binocular neurons in the frog optic tectum, *Exp. Neurol., 24,* 475–486, 1969.

Fite, K. V. and F. Scalia, Central visual pathways in the frog, in *The Amphibian Visual System: A Multidisciplinary Approach,* edited by K. Fite, pp. 87–118, Academic Press, New York, 1976.

Frisby, J. P. and J. E. W. Mayhew, Spatial frequency tuned channels: implications for structure and function from psychophysical and computational studies of stereopsis, *Phil. Trans. R. Soc. Lond. B, 290,* 95–116, 1980.

Gaillard, F. and G. Galand, A possible neurophysiological basis for depth perception in frogs: existence of a horopter surface, *J. Physiol., Paris, 76,* 123–127, 1980.

Georgopoulis, A. P., R. Caminiti, J. F. Kalaska, and J. T. Massey, Spatial coding of movement: a hypothesis concerning the coding of movement direction by motor cortical populations, in *Neural Coding of Motor Performance,* edited by W. Masson, J. Paillard, W. Schultz, and M. Wiesendanger, pp. 327–336, Springer Verlag, Berlin, 1983.

Glasser, S. and D. J. Ingle, The nucleus isthmus as a relay station in the ipsilateral visual projection to the frog's optic tectum, *Brain Res., 159,* 214–218, 1978.

Grobstein, P. and C. Comer, The nucleus isthmi as an intertectal relay for the ipsilateral oculo-tectal projection in the frog, *Rana pipiens, J. Comp. Neurol., 217,* 54–74, 1983.

Grobstein, P., C. Comer, M. Hollyday, and S. M. Archer, A crossed isthmo-tectal projection in *Rana pipiens* and its involvement in the ipsilateral visuotectal projection, *Brain Res., 156,* 117–123, 1978.

Grobstein, P., C. Comer, and S. Kostyk, The potential binocular field and its tectal representation in *Rana pipiens, J. Comp. Neurol., 190,* 175–185, 1980.

Grobstein, P., C. Comer, and S. Kostyk, Frog prey capture behavior: between sensory maps and directed motor output, in *Advances in Vertebrate Neuroethology,* edited by J. Ewert, R. Capranica, and D. Ingle, pp. 331–347, Plenum Press, London, 1983.

Gruberg, E. R. and J. Y. Lettvin, Anatomy and physiology of a binocular system in the frog *Rana pipiens, Brain Res., 192,* 313–325, 1980.

Gruberg, E. R. and S. B. Udin, Topographic projections between the nucleus isthmi and the tectum of the frog *Rana pipiens*, *J. Comp. Neurol.*, *179*, 487–500, 1978.

Grüsser, O. and U. Grüsser-Cornehls, Neurophysiology of the anuran visual system, in *Frog Neurobiology: A Handbook*, edited by R. Linás and W. Precht, pp. 297–385, Springer-Verlag, Berlin, 1976.

Hirai, Y. and K. Fukushima, An inference upon the neural network finding binocular correspondence, *Biol. Cybernetics*, *31*, 209–217, 1978.

Ingle, D., Visuomotor functions of the frog optic tectum, *Brain, Behav. Evol.*, *3*, 57–71, 1970.

Ingle, D., Spatial vision in anurans, in *The Amphibian Visual System: A Multidisciplinary Approach*, edited by K. Fite, pp. 119–140, Academic Press, New York, 1976.

Ingle, D., Detection of stationary objects by frogs (*Rana pipiens*) after ablation of the optic tectum, *J. Comp. Physiol. Psychol.*, *391*, 1359–1364, 1977.

Ingle, D., Some effects of pretectum lesions on the frog's detection of stationary objects, *Behav. Brain Res.*, *1*, 139–163, 1980.

Ingle, D., The organization of visuomotor behaviors in vertebrates, in *The Analysis of Visual Behavior*, edited by D. Ingle, M. Goodale, and R. Mansfield, pp. 67–109, MIT Press, Cambridge, 1982.

Ingle, D., Visual mechanisms of optic tectum and pretectum related to stimulus localization in frogs and toads, in *Advances in Vertebrate Neuroethology*, edited by J. Ewert, R. Capranica, and D. Ingle, pp. 177–226, Plenum Press, London, 1983.

Jordan, M., G. Luthardt, C. Meyer-Naujoks, and G. Roth, The role of eye accommodation in the depth perception of common toads, *Z. Naturforsch.*, *35c*, 851–852, 1980.

Julesz, B., *Foundations of Cyclopean Perception*, Univ. of Chicago Press, Chicago, 1971.

Lara, R., M. A. Arbib, and A. S. Cromarty, The role of the tectal column in facilitation of amphibian prey-catching behavior, *Biol. Cybernetics*, *44*, 185–196, 1982.

Lara, R., M. Carmona, F. Daza, and A. Cruz, A global model of the neural mechanisms responsible for visuomotor coordination in toads, in

Proceedings of the Second Workshop on Visuomotor Coordination in Frog and Toad: Models and Experiments, 1983, Technical Report 83-19, Computer and Information Science Dept., Univ. of Massachusetts, Amherst.

Lawton, D. T., *Processing Dynamic Image Sequences from a Moving Sensor,* PhD thesis, Univ. of Massachusetts, Amherst, 1984.

Lettvin, J. Y., H. R. Maturana, W. S. McCulloch, and W. H. Pitts, What the frog's eye tells the frog's brain, *Proc. Inst. Elect. Engrs., 47,* 1940–1951, 1959.

Ligthart, G. and C. A. Groen, A comparison of different autofocus algorithms, in *Proc. 6th Int. Conf. on Pattern Recognition,* pp. 597–600, October 1982.

Lock, A. and T. Collett, A toad's devious approach to its prey — a study of some complex uses of depth vision, *J. Comp. Physiol., 131,* 179–189, 1979.

Lock, A. and T. Collett, The three-dimensional world of a toad, *Proc. R. Soc. Lond. B, 206,* 481–487, 1980.

Marr, D. and T. Poggio, Cooperative computation of stereo disparity, *Science, 194,* 283–287, 1976.

Marr, D. and T. Poggio, A computational theory of human stereo vision, *Proc. R. Soc. Lond. B, 204,* 301–328, 1979.

McIlwain, J. T., Lateral spread of neural excitation during microstimulation in intermediate gray layer of cat's superior colliculus, *J. Neurophysiol., 47,* 167–178, 1982.

Montgomery, N., K. Fite, M. Taylor, and L. Bengston, Neural correlates of optokinetic nystagmus in the mesencephalon of *Rana pipiens*: a functional analysis, *Brain Behav. Evol., 21,* 137–150, 1982.

Muntz, W. R. A., Microelectrode recordings from the diencephalon of the frog (*Rana pipiens*) and a blue-sensitive system, *J. Neurophysiol., 25,* 699–711, 1962.

Neisser, U., *Cognition and Reality,* W.H. Freeman and Co., San Francisco, 1976.

Nelson, J., Globality and stereoscopic fusion in binocular vision, *J. Theor. Biol., 49,* 1, 1975.

Panum, P. L., *Physiologische Untersuchungen über das Sehen mit Zwei Augen*, Homann, Kiel, 1858.

Pitts, W. H. and W. S. McCulloch, How we know universals: the perception of auditory and visual forms, *Bull. Math. Biophys.*, *9*, 127–147, 1947.

Poggio, G. F., Mechanisms of stereopsis in monkey visual cortex, *Trends in Neurosci.*, *2*, 199–201, 1979.

Prager, J. M. and M. A. Arbib, Computing the optic flow: the match algorithm and prediction, *Computer Vision, Graphics, and Image Processing*, *24*, 271–304, 1983.

Raybourn, M. S., Spatial and temporal organization of the binocular input to frog optic tectum, *Brain, Behav. Evol.*, *11*, 161–178, 1975.

Robinson, D. A., The use of control systems analysis in the neurophysiology of eye movements, *Ann. Rev. Neurosci.*, *4*, 463–503, 1981.

Rossel, S., Foveal fixation and tracking in the praying mantis, *J. Comp. Physiol.*, *139*, 307–331, 1980.

Rubinson, K., Projections of the tectum opticum of the frog, *Brain, Behav. Evol.*, *1*, 529–561, 1968.

Rubinson, K. and D. R. Colman, Designated discussion: a preliminary report on ascending thalamic afferents in *Rana pipiens*, *Brain, Behav. Evol.*, *6*, 69–74, 1972.

Scalia, F. and K. Fite, A retinotopic analysis of the central connections of the optic nerve in the frog, *J. Comp. Neurol.*, *158*, 455–478, 1974.

Schipperheyn, J. J., Respiratory eye movements and perception of stationary objects in the frog, *Acta Physiol. Pharmacol. Neerl.*, *12*, 157–159, 1963.

Selker, T., Image-based focusing, in *Robotics and Industrial Inspection*, edited by D. Casasent, pp. 96–99, SPIE, 1982.

Skarf, B. and M. Jacobson, Development of binocularly driven single units in frogs raised with asymmetrical visual stimulation, *Exp. Neurol.*, *42*, 669–686, 1974.

Trehub, A., Neuronal model for stereoscopic vision, *J. Theor. Biol.*, *71*, 479–486, 1978.

Wang, S., K. Yan, and Y. Wang, Visual field topography and binocular responses in frog's nucleus isthmi, *Scientia Sinica*, *24*, 1292–1301, 1981.

Westheimer, G. and S. P. McKee, Stereoscopic acuity for moving retinal images, *J. Opt. Soc. Am.*, *68*, 450–455, 1978.

Springer